WELCOME TO FORENSICS

W9-BPO-127

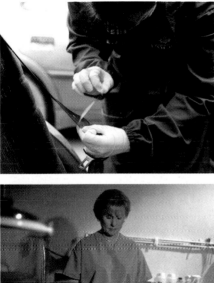

Left: A geneticist examines a DNA autoradiogram. DNA testing has become a valuable forensics tool, whether to identify unknown remains or to link a specific individual with a crime. Top: Lifting fingerprints. Investigators still rely on this century-old form of criminal identification. Bottom: A forensic pathology laboratory. The use of pathological findings hearkens back to the earliest days of forensics.

Forensics is not about one body of science. Rather, it is about how a host of sciences, and the knowledge accumulated by those who study forensics, are applied to a goal. That goal is to use science to analyze criminal evidence, and then to present the results of that analysis, as accurately and precisely defined as possible, in a court of law. The word "forensics" is of Latin derivation, from a term meaning "belonging in a forum." In the world of rhetoric, forensics describes debate or argument. In jurisprudence, it is defined as the use of science and other disciplines, such as photography or accounting, to investigate and establish evidence in criminal or civil courts of law. The two meanings of the term are linked because evidence is very much the subject of debate in the forum that is the courtroom. By necessity, a book about forensics spans the horizon of science. Few if any are the physical and biological sciences that have no forensic application in any sense. Forensics deals with subjects as varied as the timing of a rainfall and the trajectory of a single bullet. Its practitioners use tools as uncomplicated as a simple envelope to hold a fragment of evidence to a scanning electron microscope to probe the molecular structure of a piece of evidence.

Forensics has a long history—exactly how long no one is certain. What is known is that the use of forensic techniques, or what may be considered the beginnings of forensic techniques, go back several thousands of years. Thus, the first chapter of this book explores the history of forensics and how various sciences were brought into the fold of forensic usage and adapted to forensic goals. This section traces forensic history from Classical times, and even before, through the Middle Ages to the beginnings of modern forensics in the late nineteenth and early twentieth centuries.

Chapter two covers the place where forensic investigation begins, the scene of the crime, and how evidence is recovered, preserved, and passed through an all-important custodial chain that documents its travels, who possessed it, and when. Chapter three deals with the place where evidence eventually arrives for

analysis, the forensics laboratory, or—in the jargon of police work and the media that cover and portray it—the crime lab. In this chapter, readers visit major forensics laboratories and examine their resources.

Chapter four explores the foundations of forensics and one of the disciplines that hearkens back to its earliest days, pathology. It examines the role of the pathologist; autopsy procedures; and the nature of wounds and other trauma involved in accidents, homicides, and other violent crimes. In this chapter, forensic dentistry, which plays an increasingly important role in the identification of deceased individuals and of wanted criminals, is introduced.

Bones are the subject of chapter five, or, more precisely, how forensic anthropologists use bones to identify missing people and to determine circumstances surrounding deaths. Often, the bones examined by forensic anthropologists are old, even ancient. Forensic anthropologists are also called upon to investigate remains resulting from atrocities and other human-rights violations.

The contents of chapter six describe the time-tested forensic techniques of fingerprinting, ballistics, and document examinations. These methods were utilized long before more advanced forensic techniques, such as DNA testing, came into use and are still staples of criminal investigation. DNA testing, or profiling, is the subject of chapter seven. Here is the technique that has generated excitement around the world. The chapter describes the nuances of DNA testing, how it is used today, and its future

Popular interest in the work of forensic scientists has never been greater.

The unique circumstances of each crime calls for various types of photographs. An investigator familiar with the crime determines what photos will be necessary.

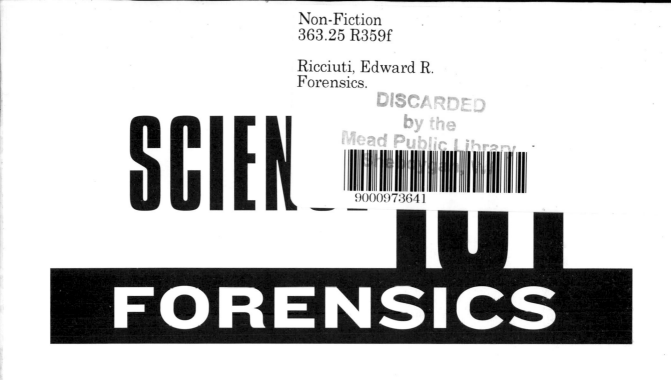

SCIENCE 101

FORENSICS

HarperCollins books may be purchased for educational, business, or sales promotional use. For information, please write: Special Markets Department, HarperCollins Publishers, 10 East 53rd Street, New York, NY 10022.

Produced for HarperCollins by:

Hydra Publishing
129 Main Street
Irvington, NY 10533
www.hylaspublishing.com

FIRST EDITION

The name of the "Smithsonian," "Smithsonian Institution," and the sunburst logo are registered trademarks of the Smithsonian Institution.

Library of Congress Cataloging-in-Publication Data

Ricciuti, Edward R.
  Science 101 : Forensics / Edward Ricciuti
      p. cm.
  Includes bibliographical references and index.
  ISBN: 978-0-06-089130-5
  ISBN-10: 0-06-089130-0
  1. Forensic sciences. 2. Criminal investigation. 3. Chemistry, Forensic.
I. Title. II. Title: Forensics

  HV8073.R552 2007
  363.25--dc22

                                    2007045479

07   08   09   10   QW   10   9   8   7   6   5   4   3   2   1

SCIENCE 101

101

FORENSICS

Edward Ricciuti

Collins

An Imprint of HarperCollins*Publishers*

CONTENTS

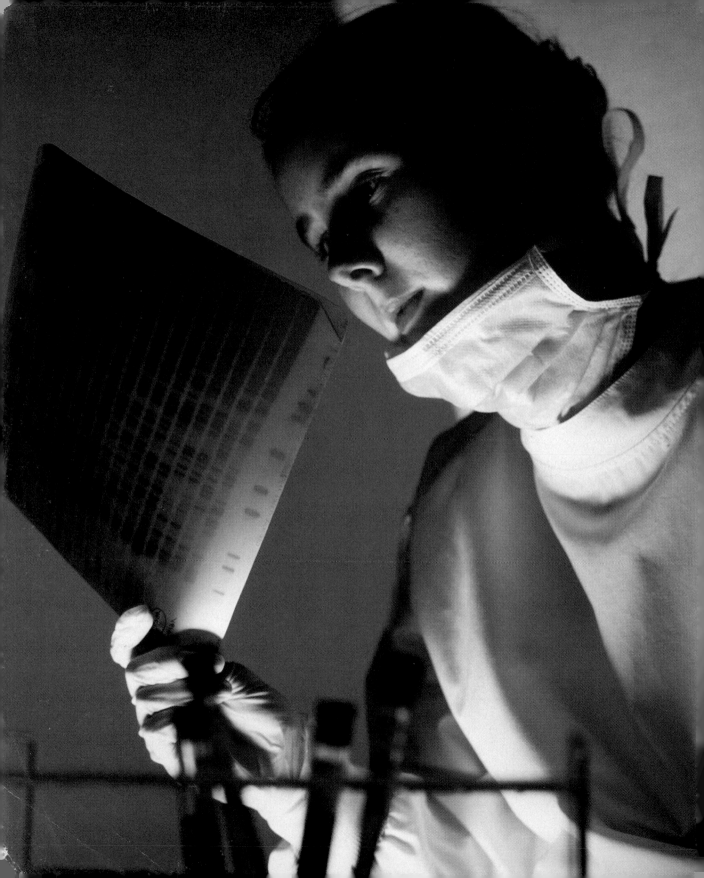

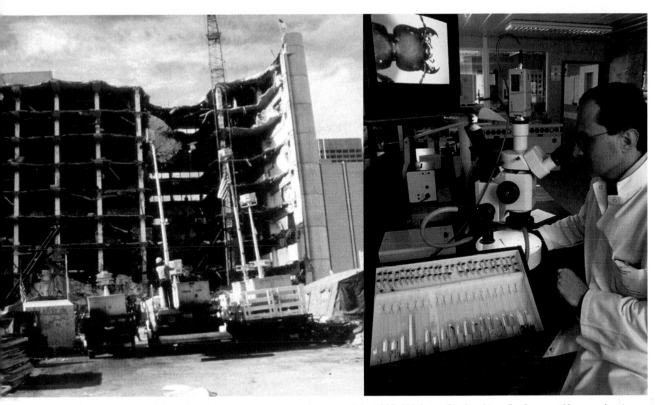

Left: The wreckage of the Alfred P. Murrah Federal Building, site of the 1995 Oklahoma City bombing. In the case of large-scale crimes such as a terrorist attack, crime scene investigators arrive on the heels of first responders, such as the police and emergency medical technicians. Right: Forensic scientists back in the laboratory analyze evidence that has been collected by on-site crime-scene investigators.

potential. Toxicology and chemistry in general are among the sciences with long histories of forensic use. Chapter eight explains how toxicologists and chemists lend their skills to investigations covering a wide range of crime, from accidents caused by drunk driving to arson and drug trafficking.

Criminal profiling—dissecting the behavior of the criminal mind—is an area that many people find especially fascinating. Chapter nine looks at criminal profiling and examines how profilers determine not only the nature of criminals but also predict what they may do in the future. Chapter ten examines how clues from the natural world are used by forensic botanists, geologists, and meteorologists to track criminals and explain the circumstances around their crimes. These clues from the earth and sky have solved many cases. Chapter eleven looks at analyzing remains to figure out how and why ancient murder victims were killed. The same forensic techniques

used to solve crimes in modern days can be used to solve those occurring in ancient times. The chapter also details human-rights cases around the world in which forensic scientists have been involved and how forensic experts help identify the victims of mass disasters, caused by humans or nature.

In a time when popular interest in the work of forensic scientists has never been greater, chapter twelve looks at how forensics in the news and in entertainment affects popular culture and the entire field of jurisprudence. The chapter surveys how forensics is presented by the media and how public perception of the legal system is based on what they see and read. *Science 101: Forensics* also contains a large section that can be used for quick reference. It contains references to key dates in the history of forensics, important forensic experts and their contributions, and many more facts about this fascinating way of fighting crime.

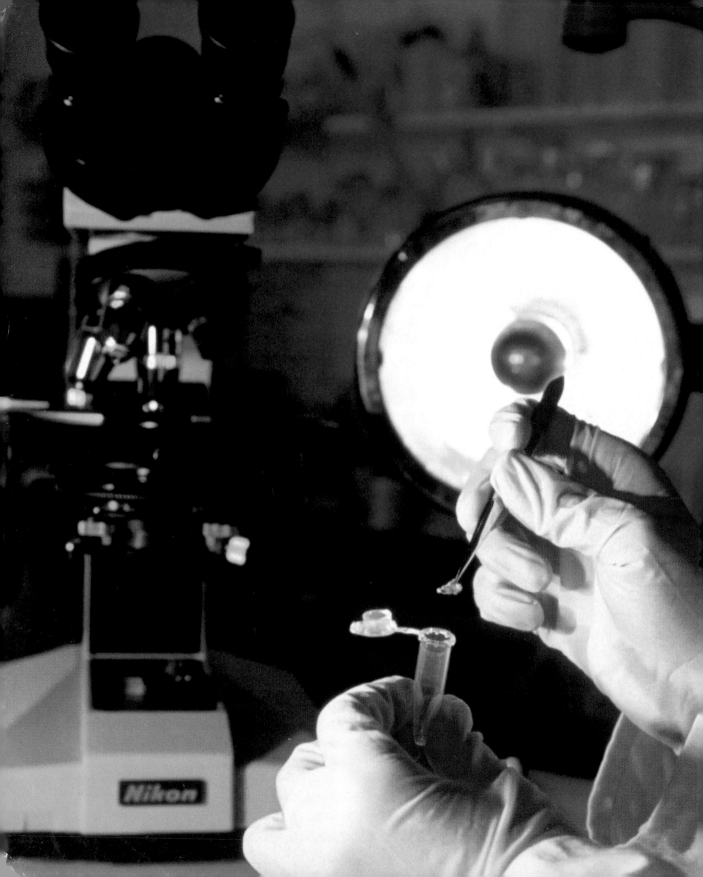

CHAPTER 1

FORENSICS: PAST TO PRESENT

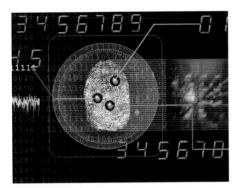

Left: A forensics scientist prepares to analyze a paint chip recovered from a crime scene.
Top: A computer visualization of scanning and analyzing a fingerprint. Identifying features of a fingerprint can be picked out by computer and matched to an individual print in a database.
Bottom: FBI laboratory technicians processing soil found on shoes for identification of clay content in the early days of the FBI lab.

Do the roots of modern forensics reach deep into prehistory, when late Paleolithic artists painted the walls of European caves with haunting impressions of their worlds, both natural and supernatural? A few of these ancient masters apparently dipped their fingertips in paint and left their prints alongside their work, as if to take credit by individual signatures.

Whether or not these early painters recognized the uniqueness of their fingerprints, there are techniques still employed in modern criminal forensics that were used far back in antiquity, although in unsophisticated form and by no means systematized. It was not until the early centuries of the second millennium that forensics began evolving into an organized body of knowledge and practice. As novel insights arose into how scientific disciplines could be applied to evaluating evidence and new technologies were invented to implement those applications, forensics developed into a force with a profound impact on the world's legal systems. Today, the pace of forensic evolution seems to be increasing geometrically as methods unheard of only a few decades ago have become routine and scientific advancement continually fosters bold new breakthroughs.

When Did Forensics Begin?

Writing in the Smithsonian Institution annual report of 1912, famed sinologist and antiquarian Berthold Laufer took issue with claims that the ancient Sumerians used fingerprinting for identification more than 4,000 years earlier. A German-born Jew who felt "healthier as a Chinese than a European," he argued that fingerprints on ancient Mesopotamian pottery were not purposeful markings, as some scholars claimed, but accidental. As might be expected from an avowed sinophile, Laufer made the argument that fingerprinting originated in China, perhaps as long as 3,000 years ago. He was probably right, if not about the dates, then about the country of origin. Although there are hints

that the ancient Babylonians may have used fingerprinting as a personal indicator, the technique seems to have truly taken hold in China, perhaps during the Tang Dynasty, which started in 618 CE.

CHINESE ORIGINS
Indeed, China can be considered the wellspring from which forensic science as a system eventually flowed. A Chinese physician, Wu Pu, supposedly settled legal cases by medical testimony in the third century. In the thirteenth century, Chinese physician and judge Song Ci wrote *Xi Yuan Ji Lu*, the classic book still studied and cited by forensics experts today. In *Xi Yuan Ji Lu*, loosely translated as "The

Washing Away of Wrongs," Song Ci outlines autopsy procedures in which he urges coroners to literally get their hands dirty, forget the stench, and meticulously examine a corpse. The book also shows how the results of autopsies could be used in court. If a victim of suffocation had water in the lungs, Song Ci explains, he died by drowning. If the throat was bruised from pressure, the cause was strangulation. These distinctions may be the first recorded instances of medical science applied to solving crime.

EUREKA!
Forensic techniques were also applied to legal issues in Europe at quite early dates. Etruscan law ordered the equivalent of a caesarean section upon the death of a pregnant woman. Quintilian, a rhetorician and attorney of first-century Rome, showed that

Left: A Roman pot. Top right: Thumbprints on ancient Chinese pottery shards. The potters sometimes "signed" their work with their fingerprints. Top left: Illustration of flies from a 1796 collection. Forensic entomology—the solving of crime by the examination of insect evidence—is one of the many forensics subfields.

a bloody palm print found at a murder scene and used to frame a man for the killing was not that of the accused. The man was freed. Quintilian's skillful use of rhetoric in his court arguments, coupled with the way he employed prints as evidence of innocence, neatly links both realms of forensics—the use of language and of science in legal situations (see introduction).

More than two centuries before Quintilian, if legend is to be believed, Archimedes used forensics of a rudimentary kind to reveal that a goldsmith had attempted to defraud King Hieron II of Syracuse. Hieron, a relative of Archimedes, asked the mathematician to determine if his new crown was pure gold, as the smith who made it claimed, or was alloyed with a cheaper metal, such as copper or silver. The bulk of the gold could not be calculated, however, without pounding or melting down the crown; it was not possible to weigh it against a chunk of pure gold of equal size. Archimedes decided to think about this problem while in the bath. As water sloshed out of the tub as he got in it, he had his famous brainstorm. He suddenly realized that his body had displaced its own volume in water—Archimedes had discovered buoyancy. Excitedly, he ran naked through the streets shouting "Eureka!" and forthwith demonstrated that the crown displaced a greater volume of water than a pure gold bar of equal weight. The crown was, indisputably, a fakery and the smith a swindler.

Early woodcut illustration of Archimedes in his bathtub. While bathing, the Greek mathematician and scientist suddenly realized that a body displaces its own volume in water. This realization allowed him to determine the weight of gold in King Hieron's crown.

An early Chinese scientist used rudimentary forensic entomology to solve a murder case in which the victim had been killed with a sickle.

TELLTALE FLIES

The thirteenth-century Chinese physician Song Ci described the first known criminal case that employed forensic entomology, although it was not called such at the time. An investigator who was asked to solve the murder of a man slashed to death by a sickle summoned all the farmers in the locale of the crime and ordered them to hand in their sickles. One of the implements—and only one—attracted a horde of flies. The investigator deduced that the insects were drawn by the faint scent of blood and tissue that still clung to the sickle blade. Confronted with the evidence, the owner of the sickle confessed to the murder.

Early Forensics

The ancient Greeks had a word for it, even if they seldom performed it: "autopsy," dissection so that a corpse could be viewed from a perspective more than skin deep. In fact, autopsy literally means to see with one's own eyes. Yet, even though they pondered the intricacies of human anatomy, the Greeks still looked askance at the dissection of dead bodies.

Still, a few accounts from antiquity suggest that anatomical examination might have been at least occasionally considered as an aid in solving crime. The Roman physician Antistius, for example, attempted to determine which of Julius Caesar's multiple stab wounds was the fatal one (the second, he said) and presented his evidence before the Roman Senate. Yet, although anecdotal evidence such as this suggests that rudimentary principles of pathology were applied to crime solving even in the ancient world, they were not applied in organized fashion until long after the Classical era. Organized application really began in the Western world during the Middle Ages. For example, judges in the courts of Charlemagne relied upon medical testimony in cases of murder, abortion, and incest. In medieval England, the title bestowed on the person invested with the authority to determine time and cause of death indicated its high status: "Coroner" derives from the Norman French word for crown, signifying a crown official. On the continent, the first recorded post-mortem examination took place in Cremona, Italy, in 1286. Meanwhile, in China, medical experts were examining bodies to determine when, if not how, they died.

MEDICAL TESTIMONY IN THE COURTROOM

Forensic pathology truly started to take form in sixteenth-century Europe when Emperor Charles V of the Holy Roman Empire ordered admission of medical testimony in homicide trials and, elsewhere, pioneering physicians began to catalog and compare how various forms of violence impacted human organs. Two names from this era are especially prominent in the annals of forensic pathology, those of Frenchman Ambroise Paré and Italian Paolo Zacchia. Paré, an army surgeon who died in 1590, was physician to four French kings and an authority on treatment of wounds from firearms. Zacchia, who was born four years before Paré died, medically attended two popes and headed the health apparatus of the Papal States.

Paré initiated numerous advances in battlefield medicine, including the revival of ligating arteries to stop bleeding after amputation. He used his battle-

Above: After examining the evidence, Roman physician Antistius opined that the second wound suffered by Julius Caesar was the fatal one. Top left: Political satire of farmers hawking over a corpse during an 1820s coroner's inquest in Britain. By the nineteenth century, the position of coroner was well known, but not always well respected.

field experience as a basis for describing the kinds of wounds caused by various weapons, taking his knowledge from the military sphere to the legal arena. Paré performed what are considered Europe's first autopsies on homicide victims and recorded these surgeries, which probably took a certain amount of fortitude, since a stigma was still attached to such activities. In 1561, he published a book describing in detail the anatomy of the human body. He explored how death by violent means affects the body's internal organs, describing consequences such as the way in which fluids fill the lungs of strangulation victims.

Zacchia, along with fellow surgeon Fortunatus Fidelis, advanced the knowledge of how disease causes anomalies in the

Holy Roman Emperor Charles V ruled that medical testimony should be used in homicide trials.

body. Considered by some of today's experts as the father of modern pathology, Zacchia bridged the gap between the pathology of diseases to pathological examination of death by unnatural causes. He may have been the first person to coin the term "legal medicine," and wrote 11 books on the subject, including *Quaestiones medico-legales*, the first compilation of expert medical opinions.

TIME OF DEATH

From the sixteenth to the early nineteenth century, Zacchia and a new wave of forensic scientists made strides in charting the progress of two conditions that

occur after death: rigor mortis, the slowly progressing stiffening of the body, and livor mortis, the process by which the body turns blue, and then purple. Their efforts made estimating the time of death far more accurate than the haphazard guesses of medieval coroners, presaging the pinpoint targeting achieved by today's forensic experts.

THE ROOTSTOCK

Pathology, most experts agree, is the rootstock from which all other forensic sciences sprouted. By the end of the eighteenth century, pathologists were joining forces at centers of forensic medicine at European universities, including Scotland's University of Edinburgh, which remains an epicenter of forensics research today. At these centers, other sciences were brought into the fold and forensics blossomed into an ever-growing field.

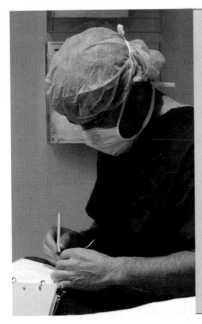

SIGNATURES OF DEATH

Rigor mortis, algor mortis, and livor mortis are changes that the body undergoes after death. The progress of their development can help determine the time at which death occurred.

Mortis is Latin for "of death." *Rigor* means "stiffness," *algor* means "coolness," and *livor* refers to an unnatural, bruised color. Rigor mortis, caused by a breakdown of chemicals in the muscles, leads to a progressive stiffening of the body that can start from minutes to a few hours after death, depending on environmental factors, such as humidity. The condition usually lasts for a day, perhaps a few hours more, and then the body gradually relaxes. Livor mortis, which usually can be seen within an hour or two of death, results when the heart stops pumping and red blood cells separate from blood serum. Cessation of the heart and the resultant shutdown of the cellular processing of oxygen cause algor mortis. By gauging the progress of these conditions and calibrating it with environmental factors at the death scene, forensic scientists can turn back the clock to the time the body stopped operating.

A pathologist must take exacting notes for an autopsy report.

Tracking the Poisoners

The poisoner's art is an ancient one, shrouded in mystery and legend. Among the poisoners of the past are figures of mythology, such as Medea, the sorceress of Colchis who tried to dispatch Theseus with the toxin of the monkshood plant, once called "the mother of poisons." Botanicals and other organic poisons favored in antiquity were largely replaced by metallic toxins by the time of the Renaissance, when poisoning was reputed to be a prime weapon of the ruthless Borgias and other ruling families of Italy. Arsenic was a poison of choice because in its white, powdered form it is odorless, tasteless, kills efficiently, and symptomatically resembles diseases. As were most other poisons of times past, moreover, arsenic was undetectable in the body.

TOXICOLOGY BEGINS

It was because poisons were so difficult to detect that they were so effective as a means of murder. As the Renaissance waned and the Age of Enlightenment began, chemistry fledged, in part from alchemy, and poisoners could no longer assume that their deeds would go undetected.

At the dawn of the eighteenth century, Dutch chemist, physician, and philosopher Hermann Boerhaave discovered that

Above: The beauty of monkshood, at right, belies its poisonous nature. Top left: Arsenic, seen here in powder form, was a favorite of poisoners from the late Middle Ages to modern times because until the eighteenth century it was undetectable in the body.

arsenic had an identifiable odor when heated. He is credited with the first use of chemistry to identify poison, but his process did not work in a corpse. By the end of the century, identification of arsenic had become much more refined. Swedish chemist Karl Wilhelm Scheele devised a method to convert arsenic found in human organs into a gas, but large amounts of the poison were necessary and there was no way to use his findings as evidence of a crime. In 1787, Johann Daniel Metzger, in East Prussia, found that if arsenous

oxide was heated over charcoal, the arsenic coalesced as a shiny, dark coating. His work and Scheele's were followed by that of German chemist Valentin Ross, who in 1806 obtained the mirrorlike arsenic deposit after treating a victim's stomach contents with a variety of chemicals to extract a compound of arsenic. As is often the case, scientific knowledge was expanding geometrically as one researcher built upon the work of another. The foundation of a new scientific discipline, toxicology, had been laid.

Swedish chemist Karl Wilhelm Scheele.

The man who erected an edifice upon that foundation was Mathieu Orfila, a Spanish professor of medicine and chemistry who is rightfully known as the "father of toxicology." In 1813, he published the *Treatise of General Toxicology*, the definitive guide to poisons and how to test for them. Among his findings was that arsenic spreads throughout the body after ingestion. With the publication of his treatise, a new scientific discipline had been born.

THE LAFARGE AFFAIR
Within a few years, toxicology had its first test in court. A French foundry owner, Charles Lafarge, had sickened and died after, witnesses claimed, his young, discontented wife, Marie, had purchased rat poison, of which arsenic was the lethal agent.

In 1840, Marie was brought to trial for murder. Unfortunately for her, a few years earlier a new test to detect arsenic had been developed. It was called the "Marsh test," after its inventor, English chemist James Marsh. In 1832, he tinkered with the work

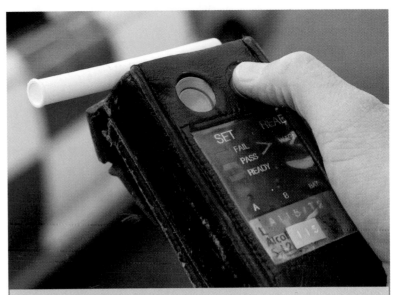

The Breathalyzer electronically measures the alcohol content of a person's breath.

LICENSE AND REGISTRATION, PLEASE
"One single drug, alcohol, has had an enormous influence on the development of forensic science." So noted Dr. Douglas M. Lucas, retired director of Ontario's Centre of Forensic Sciences, in a speech to the annual meeting of the American Society of Crime Laboratory Directors, of which he served as president. It is sometimes forgotten that the detection of alcohol in drivers belongs to the realm of forensics.

A pioneer in the field blood-alcohol analysis was Swedish toxicologist Dr. Erik M. P. Widmark (1889–1945), who in the early 1920s developed a procedure to test for blood-alcohol levels by pricking a tiny amount of blood from the fingertip or earlobe. Widmark's work led to the establishment of legal limits on alcohol in the blood of drivers, and forms the basis of the effort to stop drunk driving today. His process was later supplanted by electronic devices that analyze alcohol in the breath, such as the Drunkometer in 1937 and the Breathalyzer in 1954.

of chemists before him and came up with a way to detect arsenic within other materials even in small amounts—and to produce the shiny, black deposit that was indicative of its presence.

The Marsh test revealed arsenic in food at the Lafarge home but not in Charles's body. Orfila was called upon to redo the test and to present his findings in court. Orfila's test showed positive for the body. Another test showed no arsenic present in the ground where Charles had been buried before exhumation, blunting the defense argument that it had leached into the corpse from the earth. Marie was convicted of murder and received a death sentence, which later was commuted to a life term in prison.

Modern Forensics Arrives

In the late 1870s, Henry Faulds, a Tokyo-based Scottish physician, had become intrigued by the patterns of fingerprints he had discerned on ancient pottery. Were these marks unique to each individual potter? Faulds then studied the fingerprints of nearly everyone he knew, examining the finger ridges directly and sketching them for his records. Eventually, he developed a technique of inking each fingertip and recording its impression on paper. His studies revealed that fingerprints were indeed unique. Would this knowledge have practical applications? In 1880, assuming that fingerprints might be of use to anthropologists, he contacted Charles Darwin, hoping to secure his support in spreading this knowledge. By then old and ill, Darwin declined to help directly but forwarded Faulds's findings to his cousin, noted scientist Francis Galton. Galton never wrote back to Faulds, but nonetheless, eight months after writing to Darwin, Faulds published his work on fingerprints in the prestigious scientific journal *Nature*. Realizing that fingerprints could be used to identify criminals, he also wrote to the chiefs of the police forces in New York, London, Paris, and other major cities. He received no replies.

Meanwhile, Galton further pursued the study of fingerprints, successfully creating a basic method of fingerprint classification based on grouping the unchanging patterns of whorls, arches, and loops unique to every individual. By 1888, publication of Galton's early findings resulted in the first conviction of a criminal by fingerprinting. Using a bloody fingerprint found near a door, Juan Vucetich, a Croatian-born anthropologist and detective who had relocated to Argentina, proved that a mother who had cut her own throat had also murdered her two children.

TELLTALE PRINTS

Galton still receives the lion's share of credit for modern fingerprinting mainly because of his classic 1892 book, *Finger Prints*. As in other sciences, however, major achievements in forensics are built of work by many minds, often spanning centuries. For example, in 1686, Italian Marcello Malphigi first described lines on fingertip skin; in 1823, Prussian professor Jan Evangelista Purkinje described nine pattern classes formed by these lines (but did not suggest their use for identification). As Henry Faulds worked out his ideas on fingerprints, a British civil servant in India, Sir William Herschel, independently suspecting their uniqueness, ordered them to be affixed as signatures to contracts. After Galton's work was publicized, another British official in India, Sir Edward Henry, devised a system to further group

Left: Francis Galton used a copy of his own fingerprints in his classic book Finger Prints. *Top left: This style of "hollow point" bullet expands on impact, causing massive damage.*

each individual's print patterns so that they could be easily indexed and retrieved. In 1901, Scotland Yard adopted the "Galton-Henry system," and the practice quickly spread worldwide.

Fingerprinting replaced an earlier system of identification known as "anthropometry." Instituted in 1882 by French law enforcement officer Alphonse Bertillon, anthropometry was based on measuring and recording different parts of a subject's body in order to identify him in the future. Although this system proved useful in some cases—such as the capture of an anarchist who targeted European royalty—it had major faults. The dimensions of a person's anatomy, for instance, can change with age. Despite anthropometry's short shelf life, Bertillon is nonetheless considered the father of forensic identification for his efforts to systematize it.

THE BEGINNINGS OF FORENSIC BALLISTICS

During the period that fingerprinting gained favor in crime fighting, so did forensic ballistics. The discipline began in 1889 when a forensics professor in Lyons, France, Alexandre Lacassagne, recognized that the grooves and other deformations left in a spent bullet were made when passing through the rifled barrel.

Other researchers followed up with more sophisticated techniques, including microphotography, which revealed that each firearm left its own subtle signature of deformations on bullets discharged from it. The process was refined even further when, during the 1920s, physician and former U.S. Army colonel Calvin Goddard used a comparison microscope to match the markings on a bullet from a crime with one fired for test purposes. Today, police use the same method to trace a bullet to a particular gun. The first major case in which comparison of bullets was used in court was the trial that led to the controversial 1927 execution of anarchists Nicola Sacco and Bartolomeo Vanzetti, convicted of killing two payroll workers near Boston. Goddard also initiated a database of comparisons between different models of firearms and bullets fired from them. Among Goddard's early achievements was matching bullets to .45-caliber machine guns used in Chicago's infamous Saint Valentine's Day Massacre in 1929.

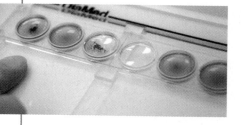

Science Tracks the Criminals

It does not take a scientist to determine that blood splattered all about can be evidence of violent crime, but it was not until the beginning of the twentieth century that law enforcement began to develop means to accurately connect blood to specific victims and perpetrators. As these techniques were refined, science became a powerful weapon used by law enforcement to track down criminals.

IN THE BLOOD

In 1901, Viennese physician and biologist Karl Landsteiner concluded after lengthy research that blood could be divided into four groups: A, B, O, and AB, as he labeled them. Thus, blood could be linked to an individual—although not exclusively. One of the first uses of blood typing in legal matters was in paternity tests.

Landsteiner's blood groups could be not identified, however, from dried bloodstains. That problem was solved in 1915 by Leone Lattes, a professor of forensic medicine in Italy, who used a saline solution to convert dried blood into a liquid form that was suitable for testing. Lattes had found small stains on his clothing and wanted to test them for blood type to determine if it was his own. It turned out to be his, caused by a problem with his prostate.

TELLTALE TRACES

As an increasing number of scientific disciplines were being applied to police work during the early years of the twentieth century, Austrian lawyer and professor Hans Gross was striving to assemble the various pieces into a cohesive system of investigating crime. Year after year, he pored over book after book, studying diverse sciences until he had organized coherent principles. In *Criminal Investigation*, published in German in 1893 and many other languages since, Gross outlined the investigatory

roles of physical and biological sciences—including toxicology, geology, botany, and zoology— ushering in the modern age of criminal forensics.

Gross particularly emphasized the importance of collecting physical evidence at a crime scene. He deemed it more important than the testimony of witnesses, noting that analysis of a smidgen of dirt on a shoe can reveal more about a crime than hours of interrogation. In effect, Gross established the modern procedure for the recovery of evidence at a crime scene, stressing that it must be a

Above: Viennese physician and biologist Karl Landsteiner first divided blood into distinct groups. Top left: Blood types can be identified by the antibodies they produce when reacting to antigens. Blood samples are shown here.

A simple shoeprint left on a sidewalk leaves many clues for a skilled forensic scientist. For example, the print can reveal the make and manufacturer of the shoe, and wear marks can even link it to an individual shoe, and hence, link its owner to the crime.

step-by-step, sequential process that is logical, rigorous, and disciplined, as are other processes of science. Advising that "all circumstances of the crime must be clearly taken into account," he urged investigators to leave no stone unturned, in that even the least traces of evidence could prove critical.

Following up on the work of Gross, French pathologist Edmond Locard focused in on the importance of detecting and collecting trace evidence, which includes even minute bits of material such as a hair or even a speck of dust or powder. Locard viewed crime scene reconstruction as a means of re-creating not only the events surrounding a crime but also the criminal himself. He likened trace evidence left behind by a criminal to the artifacts from which archaeologists re-create the lives of peoples of the past.

Locard's engrossment with trace evidence led to the formulation of a principle that, like so many revolutionary ideas, was so apparent it had never before been articulated. He never actually set it down in words, but what became known as "Locard's Exchange Principle" is simple: A criminal leaves something at a crime scene and, in turn, takes something away from it. Countless criminals have been brought to justice because of that one basic concept.

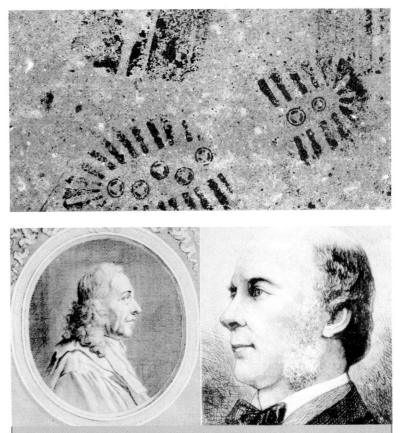

Italian physiologist Marcello Malpighi (left) was one of the first scientists to utilize the microscope to study anatomy. Sir Francis Galton (right) not only pioneered the use of fingerprinting but was also an accomplished scientist in many fields.

THE PIONEERS: MEN OF MANY TALENTS

Many of the early discoveries critical in the development of forensics were made not by men who considered themselves forensic scientists, but by those who specialized in many other scientific areas. Forensic breakthroughs were often offshoots of diverse lines of research.

Karl Landsteiner, for example, made many strides in the study of blood, including the discovery of the Rh factor in 1940, three years before he died. Francis Galton, who developed fingerprint classification, was the Leonardo da Vinci of his day. He was an African explorer, anthropologist, eugenicist, meteorologist, geographer, and statistician. Among his achievements: founding the biometric approach to genetics, the use of weather maps, and the development of differential psychology. Marcello Malpighi, whose study of fingertip skin patterns contributed to the eventual development of fingerprinting, is generally recognized as the founder of comparative physiology and using the microscope to study anatomy. He was the first person to view capillaries.

Early Crime Laboratories

The flowering of forensic science at the start of the twentieth century quickly led to the establishment of the first laboratories in which evidence was analyzed by all scientific means and equipment then available. It seems only natural that French forensics authority Edmond Locard, who emphasized the need to collect every scrap of evidence at crime scenes, should establish a laboratory solely for the purposes of analyzing materials recovered from them. The facility established by Locard, set up in 1910 within two rooms of a building in Lyons, France, was the first recognized crime lab. It was followed by the creation of others in several countries. North America's first was established in Montreal, Canada, four years after Locard's opened. Its design and operation were based on the Lyons laboratory. In 1916, the first such laboratory in the United States opened in Los Angeles, California, founded by Chief of Police August Vollmer, recognized as one of the nation's most respected police officials, despite having only a grade school education. His law enforcement career began in 1905 when through political connections he became the town marshal of Berkeley, California. While there, he founded a small forensics school, which is now a major institute devoted to the field at the University of California, Berkeley.

HOOVER'S PASSION

Advances in laboratory techniques aimed at solving crime did not escape the notice of J. Edgar Hoover, who had been appointed director of the United States Bureau of Investigation—antecedent of the FBI—in 1924. FBI historical documents describe Hoover as "captivated" by "the latest developments in scientific crime detection." Hoover's passion for forensics was noted by an adoring reporter who wrote that "the progressive director of the bureau, J. Edgar Hoover . . . is a thorough believer in science as a formidable weapon against crime."

Hoover was at first concerned with utilizing fingerprints as an identification tool, but he rapidly became aware that his agents needed the resources of other sciences as well. With the help of ballistics expert Calvin Goddard, who had established a small firearms laboratory in Chicago, Hoover installed his bureau's Technical Laboratory in Washington, D.C.'s Southern Railway Building in 1932.

Above: J. Edgar Hoover, here in 1924 shortly after being named FBI director. Top left: The University of California at Berkeley was the site of one of the earliest forensics labs.

THE BUREAU'S SCIENTIST

During its first year, the laboratory handled about a thousand examinations, mostly involving handwriting and firearms analysis. When the laboratory was founded, the bureau's only trained scientist was Special Agent Charles Appel, a stickler for details. Appel, in fact, was instrumental in its establishment, which he oversaw for his director. When the facility opened, its spartan equipment consisted of an ultraviolet light machine, a microscope on loan from the Bausch & Lomb optical company, and a wire-tapping kit. Appel had to borrow a surplus red carpet from another agency office but was able to order custom-made cabinets to hold equipment.

Appel enthusiastically took on his assignment, quickly tackling major cases. One of his first was an international sensation: the kidnapping and murder of Charles Lindbergh's son. It was under Appel's watch that the laboratory and outside experts, notably Albert S. Osborn, used handwriting analysis to identify the author of the Lindbergh ransom note as Bruno Hauptmann, found to be the killer.

Gradually, additional agents were trained for the laboratory and outside experts consulted. Two years after its founding, the laboratory was relocated to the Department of Justice Building. It became the country's—and eventually the world's—premier crime forensics laboratory.

FBI Special Agent Charles Appel, a scrupulous investigator, was the FBI laboratory's first scientist.

SCIENCE STRIKES

A month before Christmas 1933, Harrington Fitzgerald Jr., a patient in a Pennsylvania mental hospital, ate from a box of chocolates he had received as a present. He did not know the "Bertha" who sent it but was apparently pleased with the gift. Shortly after he popped the first candy into his mouth, he died of poisoning.

The candy, along with its gift wrapping and card, was sent to the laboratory operated by Special Agent Charles Appel. He compared "Bertha's" handwriting with that of Sarah Hobart, Fitzgerald's sister, who was the prime suspect. There was no match. Appel then dispatched agents to take possession of a typewriter that Hobart had sent to a repair shop. The sample of type from the machine matched that on the gift label. Confronted, Hobart confessed, and her husband was implicated as "Bertha."

A team of researchers at work in the FBI crime laboratory serology unit during its early years. Compared to high-tech facilities of today, early crime laboratories were spartan.

CHAPTER 2

SCENE OF THE CRIME

Left: Evidence recovery at the World Trade Center was a combined effort, including members of the NYPD, FDNY, FEMA, and FBI. Top: Police and rescue dogs search through the rubble at Ground Zero. FBI's Evidence Response Teams (ERTs), whose members are highly trained and motivated, were also on hand to sift through the remains. Bottom: A police officer labels vials of evidence—in this case swabs for DNA matching—recovered at a crime scene. He wears latex gloves to prevent contamination of the scene and of evidence, and to protect himself.

The scene at which a crime has been committed can be as immense as the burning towers of the World Trade Center or as small as a just-robbed mom-and-pop grocery store. But whatever the size of the crime scene, the effort to uncover what happened and who is responsible should begin at least as fast as the dust settles, with forensic examination and evidence recovery.

Before evidence can be analyzed, it must be gathered and noted in a deliberate and meticulous manner. Every scrap must be preserved and packaged, and then transported intact to a laboratory. Crime scene investigation—also called processing the crime scene—is the first and perhaps most critical phase of solving crime with forensics. Mistakes made at the crime scene can leave flaws in the foundation of an investigation that ultimately lead to the collapse of an entire criminal case. Ideally, investigators on the ground at a crime scene are trained in forensics, but this is not always so, especially in smaller law enforcement agencies, where investigators may have virtually no forensic training. At the other end of the spectrum is the FBI's Evidence Response Team, widely recognized as the best in the business of crime scene investigation.

First Response

The FBI has an Evidence Response Team (ERT) at each of its 56 field offices. Although each team operates independently, they all operate under the same standards and procedures, and when a situation demands it, teams can be combined. ERTs are supervised and coordinated by the agency's Evidence Response Team Unit (ERTU), formed in 1993 in the FBI Laboratory Division. The ERTU provides support such as training or special equipment—sophisticated video gear, for example. ERTs range in size from 8 to 50 members, some of whom are always on call. Most team members are special agents, technicians, or other FBI personnel, but often outsiders with expertise in fields such as anthropology or dentistry are enlisted for particular investigations. The FBI also has Underwater Search and Evidence Response teams in its New York City, Miami, Los Angeles, and Washington, D.C., field offices.

Agents assigned to ERTs undergo a two-week basic crime scene course taught by the ERTU. They study subjects such as crime scene management, search techniques, and documentation of evidence. Throughout their careers, agents also go through advanced training, honing the finer points of arson investigation, forensic anthropology, bloodstain patterns, and other sophisticated skills. But their jobs do not end at the crime scene. ERTs must ensure that their investigations produce legally admissible evidence, and are often called to the stand in court to explain to juries how evidence was gathered and processed.

In addition to working on land, ERTs with special training are called to specific crime scenes as needed, such as divers to perform underwater searches.

CASES IN THE HEADLINES

ERTs are usually deployed to major cases over which the FBI has jurisdiction. They were called to the scenes of the 9/11 attacks, the explosion of TWA Flight 800 in New York, the Unabomber investigation, and the bombing of the Alfred P. Murrah Federal Building in Oklahoma City. ERTs also assist state and local law enforcement agencies and play important

Above: An agent from the Federal Bureau of Alcohol, Tobacco and Firearms examines debris at the scene of the 1995 mail bomb murder of Gilbert Murray, a timber lobbyist, during the Unabomber investigation. Top left: "Stringing the tape" sets off the perimeter of a crime scene.

international roles. They were on the scenes of U.S. embassy bombings in Nairobi, Kenya, and Dar es Salaam, Tanzania, and provided extensive assistance to the U.S. government during the investigation of mass graves in Kosovo, after the conflict in the former Yugoslavia.

ARRIVING ON THE SCENE

Crime scene investigators arrive on the heels of emergency medical technicians (EMTs), police, and other first responders. Beforehand, or even on the way to a scene, agents prepare by discussing individual assignments, the kinds of conditions and evidence to be encountered at the scene, and security needs. The number of people involved in crime scene investigations can vary. The FBI's ERT usually consists of up to eight members: a leader, a photographer, a sketch preparer, an evidence collector, an evidence log recorder, an evidence custodian, and one or two expert specialists.

The first priority an ERT has on arrival is to secure the scene, to get it under control so that personnel are safe and evidence protected. In disasters such as the 9/11 Pentagon attack, it can be a massive job of establishing order amid chaos. Investigators themselves must be secure from threats such as perpetrators still in the vicinity, biological hazards, or unruly crowds. It is no easy task to secure even a small scene, because firemen, ambulance staffers, and even police milling about may corrupt evidence—by stepping on it, for example. It

While United States Marines provide security, investigators from the Canadian Mounted Police Forensics Team painstakingly sift through human remains recovered from a mass grave site in Kosovo during July 1999.

UNEARTHING AN ATROCITY

In 1999, the International Criminal Tribunal for the former Yugoslavia (ICTY) requested FBI assistance to help make its case against former Yugoslavian President Slobodan Milosevic for war crimes in Kosovo. Sixty-five ERT members, including special agents and support scientists, arrived toting more than 100,000 pounds of equipment. Their mission: to document and collect evidence at sites where civilians had reportedly been massacred. FBI director Louis J. Freeh had warned his investigators to expect "the largest crime scene in history." Even with this warning, according to an FBI report, agents were stunned by the evidence of atrocities they uncovered. One agent said "the scope and magnitude of violence we found in Kosovo defies description." The FBI forensics experts processed nine sites, including dark alleys and burned houses. They found human remains, along with spent cartridge cases and bullets. On-site autopsies were performed and witnesses interviewed. One woman described how a half dozen men, who "came in the middle of the night," forced her children to watch as they shot the men of her family. Agents later recovered the bodies, had them autopsied, and found that they had indeed died from multiple gunshots to the head.

is important to note if, in fact, evidence has been compromised during the initial response.

These days, the most familiar sign that a crime scene has been secured is the ubiquitous tape strung around its perimeter.

Stringing of the tape signals that the rush to reach the scene and secure it has ended. "Inside the tape," says a veteran ERT agent, "everyone slows down." There the meticulous processing of the crime scene begins.

Conducting a Crime Investigation

Crime scene investigators have been called up and assembled. They have discussed a basic approach to what lies ahead of them and have checked out their equipment, right down to the film in the camera. If special equipment, such as an off-road vehicle, is required, it has been made available. They have reached the scene and secured it. What comes next?

Based on FBI Evidence Response Team procedure, the search of a crime scene is generally broken down into 12 methodical and carefully sequenced steps, two of which—preparation and approach—are accomplished ahead of time.

Once at the scene, the agents initiate a series of highly orchestrated operations, each of which is equally important and part of a whole that will be the basis of an ongoing investigation. In some cases, an investigation will reveal that no crime had been committed after all. Such was the determination of the investigation into the crash of TWA Flight 800, which exploded in midair off Long Island in 1996. Terrorism was at first suspected, but based upon the evidence that was collected and analyzed by the FBI and the National Transportation Safety Board, electrical failure in a fuel tank was ruled to be the cause of the crash.

THE PROCESS CONTINUES

As the processing of the scene continues, a running description of the scene begins, either written down or dictated into a recording device. This narrative describes all aspects of the scene, including weather, lighting, tasks assigned to various personnel, and the location of evidence. Photographers are called in early on, with a variety of high-tech equipment. Virtually all elements of the scene and individual evidence are photographed, often from several angles. To fix its location, for example, a house may be imaged from one side of the street, then from the adjacent curb, and, finally, close up, by the door, revealing the house number.

Above: The National Transportation Safety Board's reconstruction of the fuselage of TWA Flight 800. Its crash was ruled an accident, but initially investigators examined the scene to determine if it was a crime. Top left: Tire tracks can be crucial evidence found at crime scenes.

A crime scene photographer snaps a shot of a firearm. Next to it is an evidence marker. He wears paper outerwear to prevent contamination of the scene.

Aerial photography is sometimes employed, especially at expansive scenes. Another way to document the crime scene graphically is by sketching diagrams, with measurements and locations of objects within the crime scene space.

As the groundwork of recording the overall scene is accomplished, the search for evidence begins in earnest. Understandably, evidence that is most likely to be overlooked, destroyed, or lost—a footprint in the sand, for instance—is collected first.

All evidence is marked, often with paper labels or small flags, and logged in by a recording investigator. Depending on the crime, evidence can range from large items such as a body to those as tiny and difficult to see as a partial fingerprint or strand of hair. Evidence is packaged according to its type. Powdery materials often go into plastic bags; moist evidence is stored in sealed containers. Hair is generally placed in plastic packets.

Investigators must be aware of everything surrounding them. Are towels in the bathroom wet? Do kitchen odors indicate recent cooking? Is the television on? Are clocks showing the right time? These observations may be unimportant—or they prove to be critical. They therefore must be noted. Once investigators complete a detailed search, they step back and review what they have accomplished. They check evidence and ensure it has been safely stored and then complete a final survey to make certain that they have left no loose ends. When all tasks have been done to the team leader's satisfaction, it is no longer necessary for the scene to remain secure, and law enforcement can relinquish authority over it. It is time, as law enforcement officers say, to "release the scene." The officer in charge will then make sure that the legal owner of the property receives an inventory of items removed, and makes an exacting record of when release occurred and to whom.

FIRST STEPS

Among the initial tasks of investigators is to interview first responders, noting their names and what they have done up to that point—whether they have moved furniture or a body, for example. Next, the leader of the team conducts a walk-through to survey the scene, and then assigns duties. In an ideal situation, each person has a specific assignment and sticks to it. Responsibilities may be so restrictive that during the search of a kitchen, for

example, one investigator may search only the cabinets. The agent who takes notes does only that, leaving rummaging to others. Of course, the ideal is not always possible. A detective for a small-town police department may have to multitask, which can lead to mistakes and omissions.

As investigators develop their search plan, they usually try to hypothesize what happened. An evaluation of evidence and a decision to call in outside experts are made as well. An experienced investigator may now consider a possibility that might not occur to the untrained: Has the scene been staged? What first greets the eye may be meant to divert investigators in the wrong direction, covering up what really happened.

Trace Evidence

Locard's Exchange Principle (see chapter one) states that when a criminal commits a crime he or she will leave something at the scene and take something away from it. That "something" transferred between scene and perpetrator, and vice versa, is referred to in the law enforcement community as trace evidence. Microscopic or visible to the unaided eye, trace evidence can link a person with another individual, an object, or a place, or can link two objects or places.

Famed forensic scientist Paul L. Kirk eloquently described the importance of trace evidence in his book *Crime Investigation*. "Wherever he steps, whatever he touches, whatever he leaves, even unconsciously, will serve as silent witness against him. . . . This is evidence that does not forget. . . .

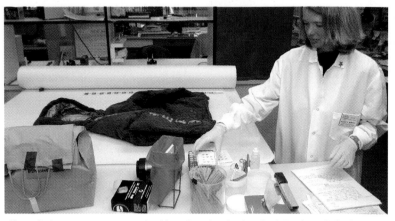

Above: Stains on a coat are examined by a forensic scientist after being brought back to the lab. Top left: A magnified section of sewing thread, which can be collected as trace evidence.

It does not perjure itself. . . . Only human failure to find it, study and understand it, can diminish its value."

The variety of trace evidence is seemingly endless: dust, semen, tool marks, fibers from clothing—indeed, just about anything can be considered evidence. Crime scene investigators must be aware of all the possibilities and understand the nuances of collecting and preserving the trace evidence that they discover.

SHREDS OF EVIDENCE

The old saw about "not a shred of evidence" implies that someone is innocent of wrongdoing or can escape its consequences scot-free. Yet, it also signifies how important even minuscule traces of evidence can be to solving crimes.

Law enforcement agencies expend considerable effort developing means of detecting trace evidence at the crime scene and perfecting those means of detection that already exist. Fibers from textiles and other materials can prove to be especially strong

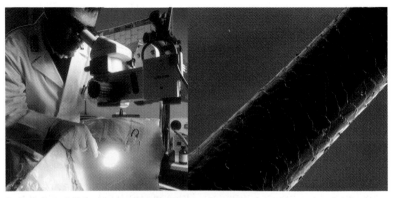

Left: A forensic laboratory technician uses a microscope to examine a sample of paint taken from an automobile at a crime scene. Right: A human hair as seen under a microscope. Hair is a common kind of evidence recovered at many kinds of crime scenes.

Crime laboratory researcher examines fluorescing fingerprints. His glasses allow him to see the glow that the fingerprint emits in reaction to the light in his hand. Organic materials, such as fingerprints, glow when exposed to fluorescent light.

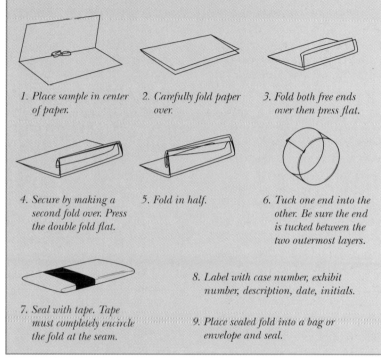
evidence linking someone to a crime. Examined microscopically in the laboratory, fibers can be compared with those of clothing, furniture, and other articles at the crime scene, or on a victim or a suspect. It is also possible to determine from fibers if an item of clothing was ripped or cut, an important consideration in sexual assault cases.

Obviously, hair from a crime scene can help determine who was there. The hair need not have been forcibly removed because, on average, a person sheds about 100 hairs each day. Hairs are generally picked up by forceps so as not to damage them.

Surprisingly, glass, plastics, and paints can be vital forms of trace evidence once analyzed in the laboratory. By piecing together glass shards, analysts can calculate the direction of the force that shattered the item to which they belonged. Since so many different items are made of plastics, they too often constitute items collected as trace evidence. Fragments of broken headlights, scraps of plastic garbage bags, strands of nylon rope—such items can be the pieces that investigators put together to picture the circumstances surrounding a crime. Repeatedly, duct tape crops up as evidence in criminal investigations. The polyethylene resin that makes duct tape sticky is essentially a plastic. Duct tape can be used to bind and gag a kidnapping or robbery victim. It has also been used in other types of crime, such as child abuse incidents in which children have been restrained by a wrapping of tape. In 2004, police found a hate message scrawled on duct tape stuck on the wall of a Jewish school targeted by arsonists.

Paint and other coatings, even tiny chips, fall within another type of trace evidence looked for by investigators. This sort of evidence can naturally provide key clues in burglaries and crimes involving vehicles—such as hit-and-run accidents—but can also figure in many other investigations, including homicides.

Blood Evidence

Evidence collected at a crime scene is usually subject to scrutiny in the laboratory by highly complex scientific equipment, but the tools used to obtain it can be as mundane and basic as items commonly found around the house. Potentially evidential dust and grit can be simply vacuumed, as long as the filter in the machine is changed with each type of evidence. Lint and fibers can be collected from hair with a comb, fingernails pared from a corpse with a nail clipper, and bloodstains scraped with a spatula or soaked up with a piece of cloth.

COLLECTING BLOOD

Blood, along with the stains, pools, spatters, and other patterns that it forms, has long been a key source of evidence, but never more so than today. In the past, the blood sampled from stains could be identified by ABO typing, which provided important clues to the identity of those involved in a crime but was not considered indisputable proof. Because blood types are not unique to individuals, linking them directly to a particular person was questionable in a court of law. DNA testing, however, can unequivocally pinpoint the source of the blood, which makes proper collecting of blood evidence ever more critical. The way in which

blood is sampled depends on whether it is wet or dry. Generally, investigators use small packets, envelopes, and bags to package bloodstains. The static properties of plastic can cause bits of dried blood to disperse and cling to the packaging, and therefore paper is

collect fingerprints. Another is to scrape the stain into a packet with a knife, paint scraper, or similar object. Cotton thread or patches moistened with distilled water can also be used to absorb blood from dry stains. Wet stains can also be packed directly or absorbed by

Above: After DNA is collected, it is chemically segmented, and then exposed on x-ray film. The segments appear as black bars; each bar represents the positions of the four DNA bases that form an individual's unique genetic code, called a "DNA fingerprint." Top left: Bloodstains are a key source of evidence, containing both DNA, and the individual's blood type.

preferred for dry stains. On the other hand, plastic serves best to protect wet blood samples from contamination.

Items with dry stains can be directly packaged, or the stain can be removed in a number of ways. One is to press the sticky side of tape onto a stain, and then lift it, similar to a technique used to

patches, and then air-dried not more than two hours after collection, in order to avoid contamination by microorganisms.

BLOODSTAIN PATTERNS

The patterns of bloodstains at a crime scene can tell at least as much about what happened as analysis of the blood itself.

Bloodstain pattern analysis is therefore a forensic specialty. When blood is present at a crime, a narrative, a sketch, and both video and still photography can document it. Photos and, particularly, videotape can show not only the patterns in which the stains were deposited but also their relationship to other objects in the area, helping investigators to map the crime reconstruction. The most basic use of bloodstain patterns is tracking down a body or perpetrator by a blood trail.

Bloodstains can be deposited passively by dripping or other actions of gravity, or projected a distance from the body by an outside force, such as a blow. Blood can also be transferred from one object to another by contact—when a bloody hand touches a doorknob, for example. Using complex mathematical formulas, forensics experts can calculate the force of the energy

A computerized, three-dimensional model of a homicide victim showing the trajectory of a bullet, indicated by an arrow directed at his chest.

that creates blood spatters. A spatter of smaller, more dispersed stains is likely to have come from a high-velocity impact, such as a bullet, while sprays of larger stains derive from an impact of lower velocity, such as a blow with a blunt instrument.

Although bloodstains are usually associated with crimes of violence or accidents, they may be related to other events as well. A burglar, for instance, may cut a finger on a broken windowpane. The smeared blood may well leave a signature on the crime.

DOCUMENTING EVIDENCE: TRUTH IN LABELING

Crime scene investigators must adhere to law enforcement's own form of truth-in-labeling laws. Logging in and labeling, or tagging, evidence is one of the most detailed tasks performed at the crime scene. All evidence must be labeled so that its integrity as evidence is maintained throughout the legal system's "chain of custody," the chronological paper trail that documents who has handled evidence and when. It is critical for law enforcement to see that there are no breaks in the chain, which would allow a legal challenge to evidence. Accurate documentation proves that evidence presented in court is the same as was found at the scene.

Labeling begins the moment an article is established as a piece of evidence. It provides a complete description of the article, who found it, the date and time it was found, and its exact location. The label will also indicate where it is to be transported. When evidence is placed in a container for transport, similar information is also noted on the outside.

A scene of the crime officer, who provides scientific support to the police, notes information about evidence, in this case a bayonet, collected at a crime scene. The investigator, wearing latex gloves, marks the evidence packaging to prevent unauthorized opening.

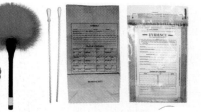

Picking Up Prints

Even in today's age of high-powered forensic techniques such as DNA testing, prints remain the most common key evidence left at crime scenes. Not only hands but also feet, shod and unshod, may leave behind their prints. Weapons and tools, moreover, leave their own distinctive traces in the form of markings. Crime scene investigators must be alert for all such forms of evidence.

FINGERPRINT RECOVERY

Forensic investigators break down fingerprints into three classes. The most common are latent prints, which are not clearly visible to the naked eye. Perspiration, oils, and other bodily moisture from the pores on fingerprint ridges, as well as those on the undersides of the feet, usually form latent prints. Patent prints, which are clearly visible to the naked eye, result when someone touches material such as paint, blood, or dust, and then touches another object. Molded prints are those from direct contact with such materials.

Fingerprints are usually analyzed in the laboratory after investigators collect the items on which they are—or may be—found. Latent prints, however, are sometimes collected

on the scene by a process called "dusting." In this process, the investigator brushes an area with a colored powder—gray, black, white, or another hue that best contrasts with the background. The grains of powder stick to the moisture in the print. Next, a clear adhesive tape is pressed down on the area, and then removed, or "lifted," taking the print impression with it. Often, investigators blow gently on the dusted surface before using tape, to remove any excess powder or air pockets among the dust grains.

SHOEPRINT RECOVERY

Impressions of shoeprints routinely remain at crime scenes, although they do not always

Above: A fingerprint shows up clearly on a knife handle. Top: Dusting for fingerprints is a tried-and-true method of gathering evidence at a crime scene. Prints are then examined at a laboratory. Top left: Fingerprint dusting material.

belong to the perpetrator. A shoe-print may be three-dimensional, as when made in mud or sand, or two-dimensional, as those imprinted on a hard surface, such as a window ledge stepped upon during a breaking and entering. Like fingerprints, shoeprints can be dusted, but several other methods of taking impressions of them are also employed. One of the most useful is a cast made of dental stone, which is much harder than plaster of paris and thus resists abrasion. A cast can show not only characteristics of the shoe sole that may help match it to the manufacturer, but also gouges and worn areas that correspond to a single, individual shoe. Electrostatic film, which attracts dust from impressions made on a dry surface, is another tool used by investigators, as is a gel mix that serves under both wet and dry conditions. Most processes for recovering shoeprints also work with those left by vehicle tires.

FIREARMS RECOVERY

When a crime involves a firearm, investigators search not only for the weapon but also for spent bullets and cartridge cases and, if a shotgun is involved, wadding and shell casings. Firearms are handled with extreme care and first checked to ensure that they are not loaded. Film and television detectives of the past were often shown picking up handguns by means of a pencil inserted in the barrel. Forensic investigators, at least those with experience, never do that. Not only is such a procedure

TOOLS OF THE TRADE

When a housebreaker forces open a window with a pry bar, the scratches and gouges made by the tool can contain markings, even of a microscopic nature, specific to its manufacture. If an object damaged by a tool, such as a windowsill, can be removed and packaged, it is sent whole to the laboratory. Otherwise, investigators make a cast with acrylic. The markings may lead to the tool, and perhaps its owner. If found, a tool can be compared to the marking it left.

This door has been broken into during commission of a crime. Tools and the methods in which they are employed, leave distinctive markings—clues to the crime.

unsafe, but it can also damage the barrel, corrupting its use as evidence. Instead, investigators gingerly lift a pistol or revolver by its handgrip, particularly if the grip is textured, as often is the case.

Above: A bullet recovered from a crime scene was deformed by impact. Right: To secure a semiautomatic pistol for transport, an investigator locks the slide in a rear position and removes the magazine while wearing gloves to prevent leaving his fingerprints.

Specialized Photography

Almost since it was invented, photography has been a key tool of crime scene investigation because it provides accurate documentation of the scene. It still is a key tool, and as new imaging technologies are developed, it becomes even more important than it was in the past. Like other techniques used to record the scene of a crime, photography includes a number of specialized areas, such as its use at night and in the infrared portion of the spectrum. Like other forensics techniques, those of photography are multiplied many times in the laboratory, but they first come into play at the crime scene.

WORKING IN THE DARK

Although highly sophisticated cameras are now readily available, many forensic photographers consider the old-fashioned, tried-and-true manually operated camera essential for photographing a crime scene in the dark. Their reasons have nothing to do with nostalgia; they prefer them because there is little or no existing light available. Nighttime crime photography relies on long exposures that cannot be attained by automatic adjustments of f-stop and shutter speed. Settings must be calculated and set manually for optimal results, and this knowledge is usually the

result of long experience and practice. Strong external flash equipment and a tripod to steady the camera are also necessary to illuminate not only the scene but also to show details so clearly that they are indisputable. Control of the scene by law enforcement is critical if the crime scene photographer is to do his or her work successfully. Extraneous lighting from streetlamps, emergency vehicles, and similar sources must be off. If the crime scene is on a highway, traffic must be stopped to reduce lighting and motion that could blur images. While the photographer shoots, other investigators in the vicinity may have to reduce or halt their own activities. Working in the dark challenges not only the skill of the photographer but also the ability of other authorities on the scene to coordinate their efforts.

Top: Student crime scene photographers photograph a dummy representing a homicide victim. Above: An arson investigator photographs evidence after a suspicious fire. Top left: Camera film is a key weapon in the fight against crime.

INFRARED-SENSITIVE PHOTOGRAPHY

A number of specialized photography techniques can detect aspects of a crime scene invisible to the human eye. Infrared photography is among those most frequently used. Infrared photography detects thermal, or heat, radiation at the lower end of the electromagnetic spectrum. This radiation has a wavelength less than that of visible light but greater than radio waves.

Because infrared photography does not depend on visible light to produce images, it is especially useful in the dark, not just at the crime scene but also for surveillance and detection of intruders. With infrared-sensitive film, photography can obtain images of objects that might not be readily apparent in a visual search. Whether in darkness or daylight, infrared also detects such evidence as a hidden body; powder burns from a gunshot; writing on paper that has been partly charred during a fire; and fibers, hair, and chemicals that otherwise might escape notice. From the air, infrared photographs can provide investigators with another perspective of the landscape of an expansive crime scene. Deciduous trees, for example, give off a higher infrared radiation than conifers.

Infrared photography is particularly attractive to law enforcement agencies because it does not involve complex processes. Budget-conscious agencies favor it because, with the right film, it can be done using almost any kind of camera.

PROCESSING A MOTOR VEHICLE

A motor vehicle can be crime scene unto itself or part of a larger one. Investigators follow a search process that is specifically designed for motor vehicles. Particular care must be exercised, for example, when crime scene investigators poke their hands around under seats or into glove compartments—a hypodermic needle or knife may be concealed in either place. Photographers follow a structured sequence, first shooting the exterior of the vehicle from several positions: sides, corners, front, and rear. Next, decals, tags, and the vehicle identification number are photographed, and then the interior, starting at the driver's compartment and ending at utility compartments such as the console and trunk.

A crime scene investigator searches an automobile, taking care to avoid any sharp objects that might be hidden under or between seats.

Trees give off heat radiation of varying intensities when photographed in infrared film.

Arson Scenes

Investigation of a fire in which arson is suspected involves many of the same procedures and personnel as at other crime scenes but differs in key aspects. Fire officials as well as police are involved, and often the fire marshal is the lead investigator. Outside experts can include electricians, electrical engineers, and, especially, chemists. Unlike a crime scene that is intact, or relatively so, fires obliterate, damage, and hide evidence, which is why they are sometimes set to cover up other crimes.

THE ROLE OF FIRE INVESTIGATORS

The basic job of investigators who arrive at a fire scene is to figure out where the fire began and then, by examining this area, determine its cause. As in other crime scenes, investigators make an initial survey of the site, noting general impressions. Because fires are usually in structures, investigators often climb above ground level, which can be risky in a crumbling building.

Trained investigators know what to look for as signs of arson. A telling clue is that equipment, files, and other valuable items were removed from a building before the fire. Once the possibility of arson arises, the search begins for an accelerant. Petroleum fuels, such as gasoline and kerosene, are among the favorites of arsonists because they are easy to obtain and use. Other accelerants

This hair dryer was found in a house destroyed by fire. Faulty electrical appliances frequently start accidental house fires, which must be distinguished from arson.

include compounds containing acetone and alcohol, as well as industrial solvents. Fires started by an accelerant generate a large amount of heat within a short period and within a fairly restricted area. A large amount of accelerant can also ignite explosively, blowing out doors and windows.

FINDING THE SOURCE

An accelerant does not necessarily vanish during a fire. Many factors influence the amount of accelerant that remains, including the material on which it was placed, the intensity of the fire, and the nature of the accelerant itself. Fires started by accelerants may burn so rapidly that they quickly exhaust available oxygen, leaving some of the fuel unburned.

Left: As soon as safety allows, fire investigators will process the scene of the blaze just as at any other crime scene. Fire scene investigation requires special expertise, and often takes place in dangerous and damaged structures. Top left: Fires always merit investigation.

Even so, accelerants may not be readily detectable. Fire investigators, however, have several tools to find them. Many accelerants fluoresce when exposed to ultraviolet light, and the longer an accelerant is exposed to heat the greater its glow. Electronic devices that detect concentrations of volatile chemical molecules in the air have long been a favorite means of finding agents that started fires. These devices are called sniffers. Since the 1980s, another type of "sniffer" has been increasingly used: dogs. Dogs, however, cannot always distinguish between similar chemicals in accelerants and in materials that have been burned.

Samples of evidence from fire scenes are packaged in much the same ways as is other evidence. Glass jars, plastic bags, packets, and the like are used. Volatile materials are sealed in airtight containers for safety reasons and because they can quickly evaporate.

An investigator with a sniffer dog searches for a fire's cause. Trained dogs can find accelerants by scent. A dog's nose is about 70 times more sensitive to smell than a human's.

BLAZING PATTERNS: HOW FIRES BURN

Most fires follow a general pattern as they burn. A fire usually burns upward and outward, so investigators check for V-shaped patterns on walls. Gases heated by a normal fire rise in a room, and then ignite materials above, so that the fire spreads vertically. The hotter a fire, the faster it rises. As a fire spreads, air currents, stairways, and barriers such as walls influence it. Burning dwindles as solid fuel or oxygen is depleted.

Smoke indicates how gases in a fire rise as heated.

CHAPTER 3

CRIME LABORATORIES

Left: The Combined DNA Index System (CODIS) blends forensic science and computer technology to enable federal, state, and local crime labs to electronically exchange and compare DNA samples. Top: A technician examines a beer can recovered as evidence under blue lighting that will accentuate any latent fingerprints. Bottom: The Integrated Automatic Fingerprint Identification System (IAFIS) allows investigators to compare fingerprint evidence with prints stored in the largest biometric database in the world.

It resembles virtually any laboratory where scientific research is performed: spectroscopes, petri dishes, and banks of test tubes. Scientists and technicians, many wearing typical white lab coats, sit or stand at benches and counters. Some peer into microscopes. Others study computer monitors. But the goal of this lab is not to find cures for diseases, produce new pharmaceuticals, or probe the secrets of the universe. It is to help track down murderers, serial killers, rapists, terrorists, and a multitude of common criminals. This is a crime lab, or, more formally, a forensics laboratory.

Crime labs vary widely in size and complexity of operations and are operated by myriad law enforcement entities, from federal agencies such as the FBI to agencies at the state, county, and municipal levels. The perception exists in some quarters that the work of scientists in a crime laboratory is glamorous, ever exciting, and packed with action, but the reality is that forensics research is like that in other sciences. It can be repetitive, routine, of excruciating detail, and intolerant of error. The course of someone's life, or even someone's life itself, may depend on it.

The FBI Laboratory

By the spring of 2003, the FBI had completed the daunting task of moving its famed crime laboratory from the agency's headquarters at the J. Edgar Hoover Building in Washington, D.C., to a brand-new facility at the U.S. Marine Corps Base in Quantico, Virginia. The laboratory, built and furnished at a cost of $155 million, is part of the complex containing the FBI Academy, long located at the base, about 35 miles from Washington.

SPACE FOR SCIENCE

Considered the best in the nation, if not in the world, the FBI Laboratory had outgrown its old facility, which shared the Hoover Building with FBI headquarters offices. As demands on the laboratory mounted and the complexity of forensic technology increased at an exponential speed, the space allotted to the laboratory became cramped and outdated. The Quantico laboratory, dedicated in April 2003, resulted from several years of planning, designing, and moving operations from the nation's capitol into a complex designed exclusively for laboratory purposes.

STATE OF THE ART

Housed in three five-story buildings, the laboratory is viewed as state-of-the-art, with public space separated from research areas. The air in laboratory areas, which occupy two-thirds of each floor, is 100-percent sterile, to reduce chances of evidence contamination. Specially designated elevators are used to transport evidence and other materials headed into laboratory sections, which provide three times more space for forensic functions and the laboratory's 650 employees than was available in the Hoover Building. Bays on the ground floor are big enough to house evidence the size of an aircraft fuselage.

FLOOD OF EVIDENCE

The FBI Laboratory processes evidence from all crimes under investigation by the agency as well as from state, local, and, sometimes, foreign law enforcement agencies. Usually, to be studied at the FBI Laboratory, cases from outside the FBI must involve violent crime, rather than offenses against property, but exceptions are made.

In an average year, more than a million pieces of evidence pass through the laboratory. Bones, paint chips, bloody patches of clothing, bullets, fingerprints, and

Above: The FBI's new state-of-the-art laboratory is based at Quantico, Virginia. Cramped space, outmoded equipment and poor evidence security sparked criticism of the laboratory when it was housed at FBI headquarters. Top left: FBI headquarters at the J. Edgar Hoover Federal Building is the former home of the FBI Laboratory.

computer files—virtually every sort of evidence imaginable comes under the scrutiny of the laboratory's staff. Annually, for example, the laboratory processes about 250,000 fingerprints through its Integrated Automatic Fingerprint Identification System (IAFIS) and more than 1,000 DNA samples through its Combined DNA Index System (CODIS). Stored in the DNA database are genetic profiles of nearly 1.5 million convicted offenders, a figure that steadily rises with the increased use of DNA tests by law enforcement. The laboratory maintains stores of thousands of products and other items against which evidence can be matched, such as a collection of 5,000 different firearms.

CODIS

Seldom out of the headlines, the use of DNA matching to help solve violent crimes has become one of the laboratory's prime functions. CODIS allows federal, state, and local crime labs to compare DNA profiles electronically. DNA profiles from crime scene evidence are matched with samples in the laboratory's data bank taken from convicted sex offenders and other violent criminals. Matches also can be made between DNA from different crime scenes, linking them together, a technique especially useful in tracking killers and rapists who strike in serial fashion. With this kind of information at their disposal, police from multiple jurisdictions can coordinate their investigations and share leads to perpetrators of serious crimes.

IAFIS operates 24-hours-a-day, every day of the year. It allows electronic response to a fingerprint inquiry within two hours, so matches can be made while a suspect is in custody.

A forensic scientist at George Washington University examines DNA samples.

CODIS AT WORK

Since it was initiated as a pilot program in 1990, CODIS has helped solve many cold cases. In October 1987, police in Prince George County, Virginia, found a woman dead after she had telephoned saying she had been raped and stabbed. Virginia forensics experts developed a DNA profile from evidence at the scene. Twelve years later, CODIS matched the profile from the crime scene to the DNA profile of a convicted rapist who had been serving time for another crime in a Virginia prison since 1989.

Inside the FBI Laboratory

More than 18 different units and teams carry out forensic investigation at the FBI Laboratory, each with its own specialty. Some of them, such as the two DNA Analysis Units and the Firearms-Toolmarks Unit, are what one might expect at any typical crime laboratory. Others, such as the Structural Design Unit and the Chemistry Unit serve functions with crime-fighting purposes that might not seem readily apparent, but which reflect the far-reaching scope of modern forensic science.

DESIGNS FOR COURT

In a sense, the Structural Design Unit does for court trials what design and exhibit departments do for zoos and natural history museums: develop displays that are reconstructions of settings in the real world. But the settings re-created by the laboratory's designers are those of crime scenes, not of natural habitats. In the FBI's wording, the unit produces "demonstrative evidence to support expert testimony during trial." This evidence includes three-dimensional scale models, exhibits, and displays that depict crime scenes, allowing jurors to obtain a clear picture of the special arrangement of objects, people, and other elements at a crime scene. The unit also has an increasingly important role in the war on terror: creating scale models of building layouts to help plan extrication of hostages.

THE NATURE OF THINGS

Materials analysis does just what its name implies. Specialists look into the nature of materials such as metals, soil, minerals, and glass that have been submitted as evidence. Examination is often at a microscopic or even more precise level, including the use of specialized spectroscopy that can identify atoms of trace elements in metal compounds.

Materials analysis has a host of forensic applications. It can suggest the reasons for a stress fracture that caused an aircraft wing to fail. Analysis of a soil sample taken from a tire or a shoe, for example, can match that sample with soil at a crime scene.

The chemical makeup of many kinds of trace evidence, ranging from building materials to paint chips to clothing fibers, often needs close analysis to ascertain their significance to an investigation. Because analysis of materials covers a

Above: To support their expert testimony before the Warren Commission that investigated the assassination of John F. Kennedy, the FBI produced this scale model of Dealey Plaza. Top left: Current FBI Director Dwight Adams leads a tour of the FBI's crime lab in 2003.

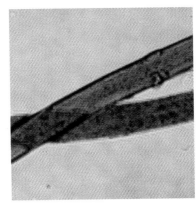

Left: Light micrographs of clothing fibers found at a crime scene, which will be compared to fiber samples taken from a suspect. If the samples match, the suspect will have been placed at the scene. Analysis of small amounts of materials can result in evidence that can make or break a case brought by prosecutors. Below left: Financial records are key parts of racketeering investigations.

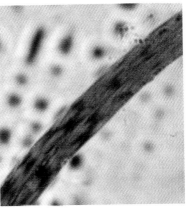

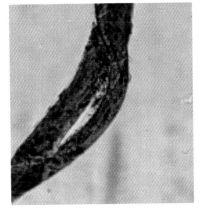

Official seal of the FBI laboratory.

FBI LABORATORY SERVICES

The FBI Laboratory provides forensic examinations, technical support, and expert witness testimony, as well as training, to federal, state, and local law enforcement agencies. It services span the gamut of the forensic sciences, including chemistry, computer analysis, DNA analysis, evidence response, explosives, firearms and toolmarks, forensic audio, video, and image analysis, forensic science research and training, hazardous materials, investigative and prosecutive graphics, questioned documents, racketeering records, latent prints, photographic analysis, structural design, and trace evidence.

broad area, it is performed in two of the laboratory's units, the Trace Evidence Unit and the Chemistry Unit.

Another type of analysis is performed in the laboratory's Racketeering Records Analysis Unit, where forensic accountants and other such specialists come into their own. This unit goes over the books, one might say, which range from the financial records of loan sharks and pimps on the street to those of executives in plush offices who profit from money laundering.

The DEA Laboratory

The Drug Enforcement Administration (DEA) is an obvious place to find a forensics laboratory, given that its job is to combat the illegal trade and use of illicit drugs, which, after all, are chemicals. Specifically, as defined in its mission statement, the DEA's role is to enforce the "controlled substances laws and regulations of the United States and bring to the criminal justice system . . . those organizations . . . involved in growing, manufacture, or distribution of controlled substances . . . in the United States."

It only follows that the DEA's forensic operations are focused largely on forensic chemistry and all its subdisciplines, although its scientific investigators also are involved in fingerprint identification

Above: Bundles of cocaine seized by DEA agents. Top left: A polarimeter lab. A polarimeter measures the optical properties of crystalline solids, such as those found in drugs.

and analysis, photography, and many other disciplines intrinsic to typical forensic entities.

INSIDE THE DRUG WAR

To a degree, the image of the DEA in the public's mind as that of bearded, shaggy-haired covert operatives going undercover to fake drug deals and hard-fisted agents smashing their way into crack houses is an accurate one. Hidden from the public eye, however, are technicians in white laboratory coats whose work with methods such as nuclear magnetic resonance spectroscopy and high-performance liquid chromatography is critical to identifying controlled substances such as heroin, cocaine, and methamphetamine, tracking

them to their origins, and monitoring what types of illegal drugs are being sold on the streets of the United States. Laboratory scientists not only help determine the types and locations of illicit drugs but also provide analyses that aid in developing intelligence to determine trends in international trafficking.

The DEA has laboratory sites scattered across the country in Dulles and Lorton, Virginia; Largo, Maryland; New York City; Chicago; Dallas; Miami; and San Francisco and Vista, California. The Chicago laboratory has a regional branch in Kansas City, Kansas; a mobile laboratory, which can be driven to crime scenes, is based in the southwestern part of the United States.

A DEA agent looks on as a fellow agent confiscates illegally grown marijuana plants.

DEA laboratories process evidence recovered not only by the agency's own officers but also by those of other federal, state, and local law enforcement entities. In a typical case, when U.S. Border Patrol agents searched a vehicle crossing from Mexico into Laredo, Texas, for example, they found brick-shaped packages of a material believed to be cocaine and similarly shaped but smaller bundles of an unknown compound. Analysis at the DEA laboratory in Dallas found that the bigger packages did indeed contain cocaine but the smaller ones were packed with harmless calcium sulfate, presumably to throw off inspectors.

THE POPPY FIELDS

Forensics scientists, including those at DEA laboratories, never know what type of evidence they will receive for analysis and how the cases involved will be resolved. When DEA agents and Kansas State Police raided the grounds of a home in central

Kansas, they found a huge expanse of poppies. The officers suspected that the plants were opium poppies (*Papaver somniferum*) and sent 10 samples of the 14,000 plants they seized to the DEA laboratory in Chicago. Indeed, when laboratory scientists crushed pods from the plants and analyzed chemicals extracted from them, they confirmed agents' suspicions. One might suspect that the homeowners would have been arrested and charged, but it turned out that no criminal intent was involved. The homeowners were unaware that, unlike other poppy varieties, opium poppies were illegal. According to the DEA, they "merely appreciated the appearance of the flowers."

Top: Dried opium poppy seedpods, photographed and measured so that they can be entered into evidence. Above: A field of poppies may look innocuous, but opium poppies are the source of all refined opiates, such as morphine, and is the source of the illegal drug heroin.

Fish and Wildlife Forensics Laboratory

The international trade in species that are protected by law because they are endangered, threatened, or otherwise rare has been likened to drug trafficking in the staggering amounts of money involved. Illicit trade in live animals, their parts, and products made from them totals about $3 billion annually, according to most informed estimates. The lead U.S. government agency for combating wildlife crime is the Fish and Wildlife Service's Division of Law Enforcement. Its agents employ stings, raids, undercover investigations, and myriad other methods shared by law enforcement in general. Until 1989, however, no crime lab anywhere in the world was dedicated entirely to solving wildlife crime. That year, the U.S. Fish and Wildlife Service opened its $4.5 billion Clark R. Bavin National Fish and Wildlife Forensics Laboratory in Oregon.

ONE OF A KIND

Named for the service's late chief of law enforcement, who pioneered covert investigation of the illicit wildlife trade, the laboratory operates just like any other crime lab. Its staff of about 330 people examines, identifies, and compares physical evidence, using myriad sciences and technologies. Their goal is to link that evidence with suspect, victim, and crime scene. Their work differs from that of their counterparts in one key respect, however, which literally makes all the difference in the world. The victims of crime investigated in this laboratory are animals of a mind-boggling number of species.

If forensic scientists in a typical laboratory sometimes have difficulties dealing with evidence from only one species, *Homo sapiens*, the problems are multiplied when more than one species is involved, such as in differentiating between caviar made of eggs of the sevruga sturgeon, a species common in the Caspian Sea, and that made of eggs of the shovelnose sturgeon of the central United States. That task faced the Fish and Wildlife laboratory after federal agents seized shipments of mislabeled caviar imported by two Maryland men. It was a complicated, convoluted case, involving attempts to smuggle illegally taken caviar, labeling shovelnose eggs as more expensive sevruga, and selling the black market caviar to a major airline, seafood suppliers, and gourmet stores. DNA testing of the eggs sorted them out, enabling agents to crack the case, which resulted in guilty pleas by the offenders.

Evidence that comes to the Fish and Wildlife lab can be fairly standard: blood and tissue samples or even an entire animal, such as the carcass of a wolf. Sometimes the evidence received is a weapon, perhaps a rifle, or even a poacher's bow and arrows. Frequently, evidence is in the form of animal parts and

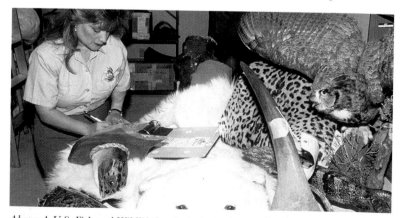

Above: A U.S. Fish and Wildlife Service inspector inventories contraband animal parts. Illegal trade in protected animals has become a worldwide issue. Top left: Fishermen deliver to authorities a haul of illegal sturgeon that had been slated for the caviar trade.

products, such as purses made of crocodile hide, pills from rhinoceros horn used in Asian medicinals, or pieces of ivory. It is this kind of evidence that presents the most difficulties to laboratory scientists. They must be able to prove in court that an item comes from a single species, not from any other in the entire world.

IDENTIFYING IVORY

By way of example, if an entire elephant was tagged as evidence, juries would have no difficulty recognizing what it was. But what about an ivory carving? Is it illegal ivory from an African elephant, or legal mammoth ivory, which sometimes enters the market? And how can the problem be solved in a way that does not destroy or damage the evidence itself?

The laboratory developed a simple way to tell the difference, using simple tools. Fine, angled lines, like stacked chevrons, show up on polished cross sections of both ivories. The lines are photocopied, then extended with pen or pencil, and their angles measured with a protractor, then averaged. If the average is above 100 degrees, it is elephant ivory; if it is below that, it is ivory from a mammoth.

Because wildlife forensics is in its infancy, researchers at the laboratory must continually seek new ways to link parts and products to individual species. Among new developments is a database whose information is based on analysis by a form of chromatography that identifies bloodstains from more than 50 different species.

Elephant tusks and rhino horns, along with a cache of weapons and ammunition, that were seized from poachers in Zambia. Although the African elephant and the black rhinoceros are both protected species, the black-market trade in their body parts has not ceased.

DEATH OF AN EAGLE

Federal agents and Kentucky conservation officers investigating the death of an adult bald eagle and 41 black vultures on the lands of a private hunting club found a container of a highly toxic pesticide and equipment to inject it when they raided the club's lodge. The agents and officers believed that the federally protected birds had died from eating poisoned bait, which had been set out to kill mammalian predators. The birds were sent to the National Fish and Wildlife Forensics Laboratory for analysis, which verified the agents' suspicions. Based on the evidence, the corporation that owned the club pleaded guilty to the violation and paid a $15,000 criminal fine.

A bald eagle flying in for a landing. Federal law protects threatened or endangered birds, such as the bald eagle and the black vulture. One of the duties of the United States Fish and Wildlife Service is to make sure that those laws are not broken, whether knowingly or unwittingly.

ATF Laboratories

With the shock waves from the World Trade Center disaster still reverberating and terrorist bombs exploding around the world, the mission of the Bureau of Alcohol, Tobacco, Firearms and Explosives (ATF), the focus, and even the name of the agency were redefined with its transfer, in 2003, from the U.S. Treasury Department to the Justice Department. When the switch was made, the word "explosives" was added to the ATF's name, though not to its acronym. Established in 1972 as an independent bureau, the ATF had always been charged with solving and preventing crimes related to explosives, as well as firearms, and arson, but that extra word indicated an increased stress on preventing bombings, if not an attempt to calm public fears. As for alcohol and tobacco, stopping illegal traffic in those items has always been the province of the ATF and of its long lineage of predecessor agencies, dating back to alcohol tax collectors in 1791 and running through the boot-legger-busting agents of Prohibition.

BEGINNING IN AN ATTIC
Forensics responsibilities in the ATF fall under its Office of Laboratory Services. Its origins began in a laboratory established by Congress in 1886—composed of two scientists ensconced in the Treasury Building attic—to examine alcohol products, as part of revenue enforcement. The office operates forensics laboratories in San Francisco and Atlanta and at its $135 million National Laboratory Center in Ammendale, Maryland, which also is the site of the ATF's Fire Research Laboratory. More than 100 scientists and technicians work at these facilities.

ATF forensics scientists work in areas similar to those of their counterparts at other crime laboratories, such as comparing trace evidence, examining fingerprints, checking the authenticity of questionable documents, and examining firearms and toolmarks. Given the mission of the ATF, documents under scrutiny may be counterfeit cigarette tax stamps or papers damaged in an arson fire. Tracing firearms is a forte of the labs. In 1994, the ATF initiated a unique computerized system for matching spent bullets with the firearms that discharged them. ATF ballistics expertise came into play during the investigations of the 2002 Maryland sniper shootings, which caused widespread panic

Above: In 1923, prohibition agents, precursors to later ATF agents, raid a lunchroom speakeasy in Washington, D.C., to stop the illegal sale of alcohol. Top left: The ATF is known for enforcing federal laws relating to alcohol, firearms, and explosives, but it also retains its role in investigating the illegal trafficking of tobacco products, such as cigars and cigarettes.

in the Washington, D.C., area. ATF ballistics experts matched spent .223-caliber bullets and shell casings from the shooting scenes to a rifle found in the car of John Allen Muhammad and Lee Boyd Malvo, critical evidence in proving their guilt.

IN CASE OF FIRE

The focus of the ATF on arson and crimes related to explosives translates to top-shelf forensics expertise in these areas. Research on explosives centers on not only determining their composition but also tracing them to their points of origin, not an inconsequential goal when combating a terrorist network that spans the globe. The agency's fire research lab brings to bear a greater number of technological tools on arson than does any other place in the world. Besides laboratory space, the laboratory has mammoth test bays, the largest of which can accommodate a two-story building. Within these bays, researchers can analyze burns ranging from those on vehicles to those of entire buildings.

The fire laboratory's role goes beyond supporting investigations by the examination of evidence. Research under way there is aimed at developing new methods of reconstructing fire scenes, which may result in new evidence-gathering techniques. This research may even result in a new list of what is even considered evidence of arson.

ATF agents investigate a burned-out chemical laboratory building in Kansas, in 2002. The blaze spread from the laboratory and damaged nearby structures as well.

The Integrated Ballistic Identification System (IBIS) is a computerized database that connects more than 200 federal, state, and local law enforcement agencies.

IBIS

Add to computerized databases of DNA characteristics and fingerprints the ATF's Integrated Ballistic Identification System (IBIS), which is linked via a network to more than 200 federal, state, and local law enforcement agencies and laboratories. Almost 900,000 pieces of evidence have been entered into the system and, by 2006, more than 11,000 matches of bullets and cartridge casings to firearms were made by police agencies using it. The assistance provided by IBIS to local agencies can be considerable. In December 2002, the ATF announced that more than six firearms laboratories in the United States had recorded more than 500 matches each. Some of the information thus obtained for investigators, the ATF noted, was unobtainable by any other means.

Laboratories for All Purposes

It comes as a surprise to many people that more than 100 federal agencies have some degree of law enforcement powers. The mission of some, such as the FBI, U.S. Secret Service, and U.S. Bureau of Customs and Border Protection (BCP), part of the Department of Homeland Security, is law enforcement. For others, including the U.S. Postal Service and Food and Drug Administration (FDA), law enforcement plays an ancillary role to their main function.

BCP LABORATORY

Many of these agencies have their own forensics laboratories. The core activities of most crime laboratories, whatever the agency, are similar, but some also specialize

in areas that reflect their particular missions. The laboratories and scientific services staff of the BCP, for example, manifest the role of U.S. Customs in enforcing trade regulations and policies. Since U.S. Customs laboratories were founded in the late 1800s, determining the countries where products originate and ascertaining that shipments truly are as labeled have been two of the labs' basic responsibilities. A shipment of bed linen labeled as cotton, for instance, may require trace analysis to determine whether sheets are 100 percent cotton or contain synthetic fibers.

With the consolidation of Customs, the Border Patrol, the Secret Service, the Coast Guard, and several other agencies into

CBP uses a mobile laboratory when needed at ports of entry around the United States.

the Department of Homeland Security (DHS) in 2002, the laboratories took on a broader role in support of protecting the nation's borders against weapons of mass destruction and interdiction of terrorists. Materials intercepted at the border and suspected of being radioactive—with dirty-bomb potential, for example—come under laboratory scrutiny.

The BCP laboratories analyze evidence gathered during criminal investigations by DHS agencies, notably Immigration and Customs Enforcement (ICE), whose agents are in the thick of counterterrorism efforts. Thus, the laboratories routinely process fingerprints, DNA samples, and other evidence typically recovered at crime scenes. BCP operates eight laboratories in Chicago;

Above: A mobile laboratory technician checks steel pipe entering at the border. Top left: The United States Border Patrol mounts horseback patrols through rugged, isolated country.

New Orleans; San Francisco; Newark, New Jersey; Springfield, Virginia; Savannah, Georgia; and San Juan, Puerto Rico.

SECRET SERVICE

The national interest sometimes dictates that an agency's special forensics expertise be expanded into nontraditional roles. The Forensic Services Division of the Secret Service has a handwriting information system that allows document examiners to scan and digitize handwriting, and then search and match those scans against previously recorded samples. In effect, this system is a national handwriting repository for comparing letters and other handwritten documents. The system had been used only for what the service calls "protective intelligence"—analyzing notes threatening the president and other high-ranking officials. In 1994, Congress mandated the agency to provide forensic support in cases of missing and exploited children. Since then, this sophisticated system is available to federal, state, and

The envelopes containing anthrax-laced letters sent to NBC's Tom Brokaw and Senate Majority Leader Tom Daschle in 2001.

THE MOUSE IN THE CAN

When the mouse arrived at the FDA's Forensic Chemistry Center, little of it remained for scientists to examine. The rodent, which a woman claimed she had found inside a can of a soft drink when she opened it, had been examined, reexamined, and dissected by the soft drink company, a veterinarian, and a pathologist. Little was left but the mouse's tiny teeth. But that was enough to prove that the claim of contamination was a scam.

Scientists at the laboratory measured the spaces between the mouse's teeth and the bite marks on the can. Based on a comparison of the measurements, they determined that the lower teeth had left marks inside the can and the upper teeth on the outside of the top, at the edge of the pull-tab opening. The evidence showed that, indeed, the mouse had been in the can—not, however, before it had been opened. How could the scientists tell? The can's top and pull-tab are a single piece throughout the manufacturing process. If the mouse had been sealed inside the can, it would not have been able to gnaw on the outside of the can top. Thus, the mouse could only have been placed in the can after it was opened. The tale of the teeth marks was used to convict the woman of violating the Federal Anti-Tampering Act, a felony.

FDA scientists monitor the integrity of products that end up on retail shelves.

The FDA's Forensic Chemistry Center is a 75,000 square-foot one-story building in a research park in northeast Cincinnati. It was established in 1989 to give the agency a team of forensic experts able to respond immediately to tampering incidents. The laboratory has developed a screening process for more than 250 poisons and, with a specialized type of mass spectrometry, can find contaminants in amounts as small as parts per trillion.

local law enforcement agencies handling such cases. Another tool the Secret Service uses to trace documents to their source is its International Ink Library, the most complete forensic collection of writing inks in the world, containing more than 7,000 samples. With it, investigators can identify the dye type and make of a writing instrument by the earliest date possible it could have been manufactured. Not surprisingly, the U.S. Postal Inspection Service forensic laboratory—in Dulles, Virginia, with satellites in New York, Chicago, and Memphis, Tennessee—also has facilities for ink analysis, along with comparisons of handwriting, type, and print. Postal Service scientists are adept at restoring writing that has been erased or obliterated.

State and Local Crime Laboratories

The rape and murder of a college coed had gone unsolved in New York State Police files for 28 years. Committed before DNA testing was developed, the crime constituted the coldest of cold cases—until, that is, a state police investigator asked the New York State Police's Forensics Investigation Laboratory system to reconsider the case. Shortly after the crime had been committed, a laboratory technician had prepared a slide with a semen sample recovered from the victim. The still-extant sample was subjected to DNA testing, and the results led to the apprehension and subsequent conviction of one of the suspects in the case.

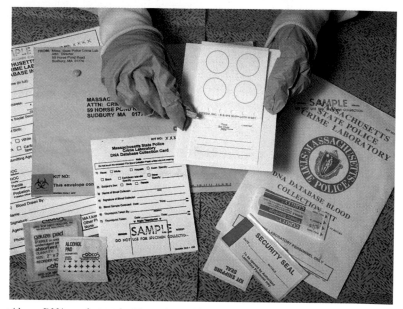

Above: DNA analyst at the Massachusetts State Police Crime Laboratory displays a kit used to collect blood samples for the state's DNA database. Top left: A DNA strand.

STATE LABORATORIES

State law enforcement agencies, as well as those of many cities and counties, have their own forensics laboratories that investigate cases within their particular jurisdictions. The New York State forensic laboratory system's main facility is at state police headquarters in Albany, with three regional laboratories distributed across the state in Newburgh, Port Crane, and Olean. Started in 1936, the laboratory, like those in other states, uses the full spectrum of DNA identification technologies and enters DNA patterns of

violent offenders into CODIS, the nationwide Combined DNA Index System.

One of the most comprehensive crime laboratories in the United States is Georgia's Bureau of Investigation. Established in 1972, it has nine sections devoted to forensic specialties ranging from trace evidence to questioned documents. In terms of space, the Arkansas State Crime Laboratory is as large as some federal facilities. Established in 1972, it occupies an 80,000-square-foot building in Little Rock. It is not part of the state's Department of Public Safety or state police, as

many state laboratories are, but a separate entity with a director appointed by the governor. The laboratory accepts evidence from law enforcement agencies within the state as well as public defenders. The crime laboratory section of the Delaware State Police differs from many others in that it is not a full-service laboratory but one that specializes in three areas: blood and breath alcohol analysis, questioned document examination, and forensic photography. In a typical year, the photography unit produces more than 26,000 photographs for law enforcement purposes.

It is sometimes necessary to hire outside help, such as professional carpet cleaners, to clean up rooms after a crime scene investigation.

ADVICE FOR HOMEOWNERS

Fingerprint powder sprinkled around a crime scene can cause a mess, and crime scene responders are sometimes less than scrupulous about cleaning it up after finishing their jobs. Here, somewhat adapted, is advice for the home-owner faced with tackling the job, courtesy of the Forensics Division of the Department of Public Safety, Greenville County, South Carolina:

- Vacuum up the powder before trying to use a wet mop. Do not let the vacuum brush drag through the powder, which can spread it and grind it into the floor.
- Wash powder off cloth surfaces with a mild detergent.
- If the above steps fail, contact a professional cleaning service.

BIG-CITY LABORATORIES

Big-city crime laboratories vary in the way their functions are assigned, depending on the organizational structure of their respective police departments. The Los Angeles Police Department's Scientific Investigation Division has a technical laboratory and criminalistics laboratory. The technical laboratory focuses on fingerprint investigations, photography, documenting crime scenes, polygraph examinations, and providing electronic surveillance devices for investigators. Work at the criminalistics facility tends to be more analytical in nature, including DNA testing, examination of trace evidence, and toxicology. The New York City Police Department's Forensic Investigation Division is more directly linked to in-line operations of the department in that it is part of the Detective Bureau. The division has a laboratory for

Shouldering one of the most dangerous of all police duties, a New York City bomb squad member suits up before investigating a suspicious package left in Times Square.

scientific examination of evidence and separate units for ballistics and fingerprint identification. The department's bomb squad, which has one of the most dangerous tasks in law enforcement, is part of the Forensic Investigation Division.

Crime Laboratories around the World

Throughout the world, it is known as "Scotland Yard," or, more familiarly, the "Yard." But odds are that a sizable proportion of people who have heard the name—outside of the United Kingdom, that is—do not know the official designation of the law enforcement agency it signifies. Scotland Yard is Britain's Metropolitan Police Service, which polices Greater London.

As a premier law enforcement agency, Scotland Yard has an extensive forensics operation, which was consolidated in 2001 from a number of metropolitan police units and services.

The work of Scotland Yard's Forensics Services ranges from following up on burglaries to antiterrorism operations and encompasses the spectrum of forensic activities, starting with

evidence recovery at crime scenes. Specialists are assigned to particularly serious crimes, such as armed robbery and homicides. Individual units are assigned to tasks such as evidence recovery and forensic photography, but DNA profiling is performed by outside organizations.

On a national scale, the Netherlands has a forensic institute within its Ministry of

Above: Scotland Yard's National Fingerprint Gallery at the Metropolitan Police Support Headquarters in London contains more than 4.4 million fingerprint records. Top left: The Netherlands Ministry of Justice Building at The Hague, which maintains a forensic laboratory.

Justice. Established in 1945, shortly after the close of World War II, the institute operates a laboratory for pathology and another for other forensic disciplines, such as digital evidence technology and DNA analysis.

CANADIAN LABORATORIES

Ontario, Canada, can claim one of the world's most extensive forensic science facilities in its Centre of Forensic Sciences, under the province's Ministry of Community Safety and Correctional Services. Headquartered in Toronto, with a regional branch in Sault Sainte Marie to serve the northern part of the province, the laboratory dates to 1951. Few, if any, crime laboratories have devoted so much organizational effort to quality control. All evidence submitted to the Ontario facility goes through a receiving office that conducts rigid screening to ascertain that guidelines for submission are met, packaging is proper, and the chain of custody is observed. Another unit monitors laboratory operations to ensure that evidence is secure and its integrity maintained. The unit screens laboratory work for quality assurance and potential flaws and recommends improvements.

The province of Quebec's crime laboratory is the oldest in North America. The Laboratoire de sciences judiciares et de médecine légale was founded in 1914, four years after the creation of the world's first, in France, and two years before the United States' first, in Los Angeles, in 1916. From its inception, the

The famous revolving sign marking the entrance to New Scotland Yard. Founded in the nineteenth century, the Yard has a worldwide reputation for excellent police work.

WHERE IS SCOTLAND YARD?

Scotland Yard is not in Scotland Yard anymore. The original headquarters of the Metropolitan Police Force was housed in a building on Great Scotland Yard, a street off Whitehall, which is a main road running from Parliament Square to Trafalgar Square in London. The police are now headquartered in a building in Westminster, only yards away from the spiky towers of the British Houses of Parliament.

The derivation of the name "Scotland Yard" is uncertain. The most popular explanation, however, states that it refers to the site of the residence once used by Scottish kings and dignitaries while visiting England.

laboratory used basic forensic techniques from medicine, chemistry, physics, and biology, and as the study of forensics has advanced, it has incorporated genetics and electronic and computer engineering in its arsenal. Traditionally, the laboratory has served police agencies, coroners, courts, and various other government agencies but has begun to offer some services to the private and quasi-public sectors.

Wilfrid Derome founded the Laboratoire de sciences judiciares et de médecine légale, the first crime lab in North America.

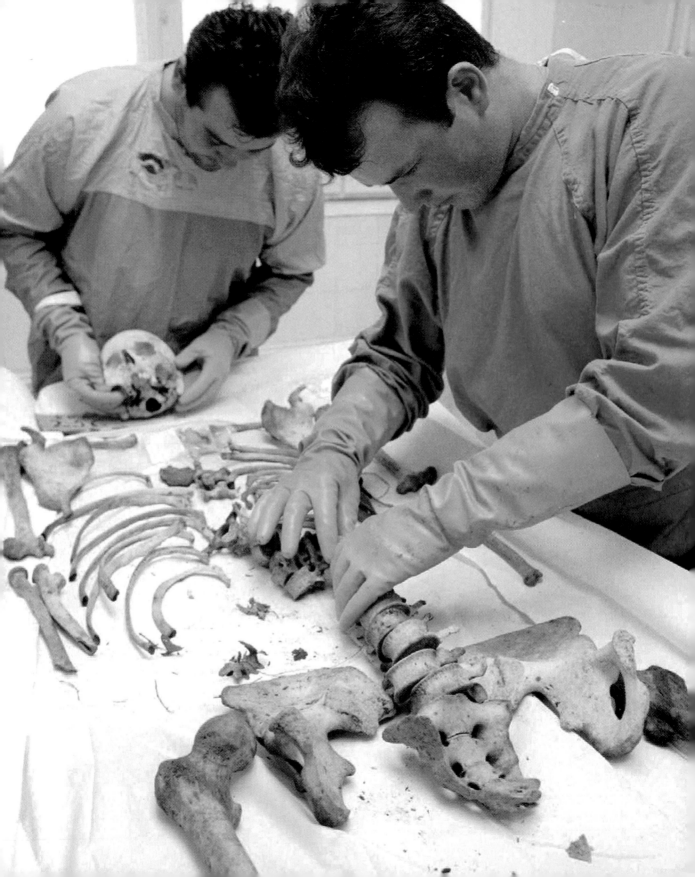

CHAPTER 4

FORENSIC PATHOLOGY

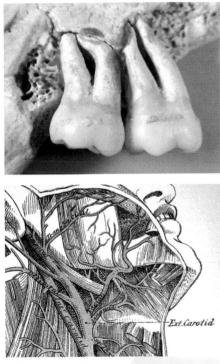

Left: Forensic pathologists study the skeletal remains of bodies found in mass graves near Srbrenica, the site of a 1995 Bosnian War massacre. These medical specialists are attempting to identify the bodies. Top: Dental pulp is a source of DNA, which can be useful in identifying victims. Bottom: An illustration from Allen's Anatomy, *an 1882 medical dictionary, would have been useful to coroners of the time, who may not have received formal medical training.*

Dead men tell no tales. That is an old myth easily debunked. Long before forensic science was established as a recognized body of knowledge, the examination of human remains already yielded a wealth of information about how homicides occurred and, often, who was responsible.

Pathology, along with the autopsy, its basic means of investigation, forms the core of forensic science. Although human dignity, the feelings of kin, and respect for the dead must be considered, the body of a homicide victim is a piece of evidence—and is treated as such by investigators. All aspects of evidence handling, such as chain of custody, integrity of storage and of samples, and recording every manner in which the evidence is examined, must be taken into account if the evidence is to have legal weight.

Dealing with the dead in a clinical manner is not for everyone. It is a skill that can require professional detachment and, in everyday language, a strong stomach. The forensic scientists who perform this task, however, are as a rule as equally passionate about solving crime as the most zealous of law enforcement investigators.

Coroner or Medical Examiner?

Sooner rather than later, the remains of a person who has died because of a criminal act or under circumstances viewed as suspicious find their way to a public servant vested with the responsibility of determining the cause and circumstances of the victim's death. That job typically is assigned to a coroner or a medical examiner, who investigates cases within a state, region, county, municipality, or other geopolitical subdivision.

Most areas depend solely on a coroner or medical examiner system, but in a few places one individual can hold both of these positions. In 1990, for example, the county of Los Angeles Board of Supervisors created a Coroner Department with a chief medical examiner–coroner at its helm.

PROBING DEATH
Whatever the office is called, the person who holds it is charged with inquiring into the death and determining what happened, usually with the help of investigators and medical experts, such as forensic pathologists and scientists who may or may not be on staff. Their goal is to determine whether

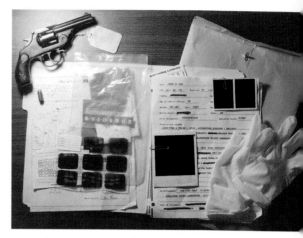

Above: An examination room, where biological evidence is documented and analyzed. Evidence such as blood, semen, or saliva can be viewed by using an alternate light source. Top left: The tools of a medical examiner.

Top: Homicide case records might include a gun, bullet, dental X-rays, an autopsy report, and gloves. Bottom: Chief Medical Examiner of Massachusetts, Dr. Richard Evans, in the state's central autopsy suite in Boston.

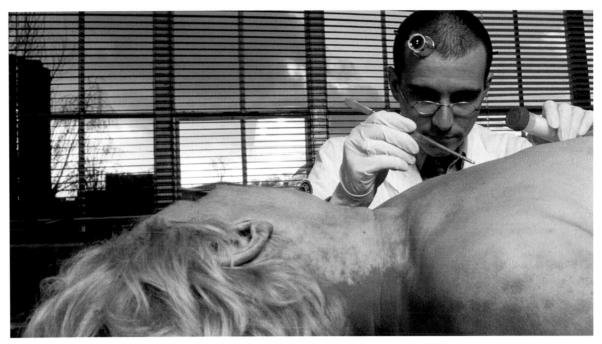

A medical examination can help estimate the time of death by studying the presence of insects, which begin to colonize a body immediately upon demise. Here, German forensic entomologist Mark Benecke examines a corpse for the presence of insect organisms such as blowflies, ham beetles, or dermestid beetles, which arrive in a particular succession.

the victim's death was natural or resulted from an accident, suicide, or homicide and, if it was accidental, whether fault, such as negligence, was involved.

SIMILAR ROLES

The key difference between a coroner and medical examiner is not the mandated function of the position but the way in which the position is performed and the qualifications necessary. A coroner, an office dating back to medieval England, is a political creature, elected or appointed, and in some jurisdictions can be a sheriff, justice of the peace, or other functionary within the legal system. Coroners need not be physicians, although some are and many undergo some medi-

cal training, on or off the job. Not infrequently, coroners are morticians who operate their own funeral businesses.

Unlike a coroner, a medical examiner—a position that evolved more than a century ago in the United States—must be a physician, although not necessarily a pathologist. The position mandates that a person who assumes it has a background in legal medicine and, in certain places, medical examiners have law enforcement powers and retain investigators on their staffs. The medical examiner's task parallels that of a coroner but can be more extensive in that it can involve personal performance of pathological examinations, such as autopsies.

Laws within particular jurisdictions govern which deaths must be reported to a coroner or medical examiner. As in other kinds of evidence, a corpse in a case investigated by the office of a coroner or medical examiner is subject to strict rules of custody and remains under the office's authority until a death certificate is signed and the body released to the next of kin. The coroner or medical examiner does not need the permission of the victim's family to order an autopsy. Physical examination of the body, however, is not the only evidence-gathering procedure that may be employed. Coroners, although generally not medical examiners, can quiz witnesses during their investigations, known as coroner's inquests.

Dead Men Tell Tales

When a medical examiner or coroner orders an investigation into a death, a forensic pathologist is assigned to the job. The task of a forensic pathologist may seem undeniably grisly to most of us because, first and foremost, it ultimately involves taking apart the human body. It is like solving a jigsaw puzzle in reverse, with the puzzle solution being the reason for someone's death.

Pathology is both a science and an art because it involves not only a deep knowledge of human anatomy but also the ability to interpret anomalies that come to light when the body is examined.

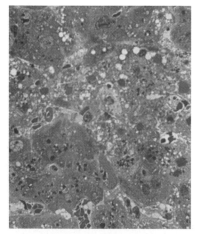

Above: A clinical pathologist may determine the cause of death by studying tissue samples taken from a body. This photomicrograph shows hepatitis caused by the Lassa virus. Top left: A tag identifies a corpse at a morgue.

Pathology is the branch of medicine that examines and interprets structural changes caused by disease or injury. It becomes a forensic discipline when an illness or injury appears unnatural or suspicious.

CASES FOR PATHOLOGISTS

Obviously, many of the cases that demand the skills of a pathologist are homicides, suicides, or traumatic accidents. Often, however, no criminal element is involved, as when an infant or an elderly person dies without apparent cause or when authorities suspect a potential threat to public health.

The pathologist approaches the body from two perspectives: anatomic and clinical. Anatomic pathology deals with the structure of the body and involves techniques such as surgical dissection and examination of organs, tissues, and even individual cells. Clinical pathology examines samples taken from the body and involves a host of laboratory disciples, including toxicology, hematology, and microbiology. Scientists from many other

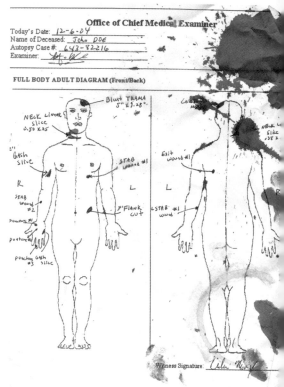

A pathologist's blood-stained autopsy report of a stabbing victim shows points of entry and exit of stab wounds.

disciplines are involved in the clinical sphere. In a case of suspected poisoning, for example, a pathologist may remove a sample from the stomach of a victim, and then send it to the laboratory for a toxicologist to examine for the presence of foreign chemicals. When called upon to identify an unknown body or skeletal remains, the pathologist can seek the expertise of a forensic odontologist to trace dental records or a forensic anthropologist to examine bones.

ANSWERING QUESTIONS

Although the main goal of the pathologist is to discover the cause of death, that discovery represents only the top of the pyramid of the investigation. For successful conclusion of a criminal investigation, myriad other questions must be addressed. Although it cannot speak, the body of a victim may answer many of these questions. The state of rigor mortis or change in body temperature can help determine the time of death. Gunshots, blows with blunt instruments, knife wounds, and other sources of trauma all leave distinctive marks on the body. The state of organs such as the lungs can indicate strangulation or drowning. Marks on the body can indicate whether it was dragged along the ground or pulled by a rope.

Forensic pathology can also play a preventative role. It can turn up the first traces that a highly contagious disease has been the cause of a death and thus allow the pathologist to alert authorities to a potential public health threat. Examination of a victim who has died from cardiac problems can reveal that the individual has a genetic defect that also may be present in other family members. A forensic pathologist can elicit volumes of information from a body that is not able to speak for itself.

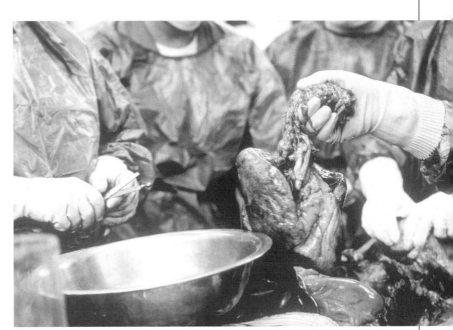

During the course of an autopsy, medical examiners remove the lungs and trachea of a 20-year-old woman, whose death was caused by a cardiac virus. Evidence from such an examination can help further knowledge about a disease or can uncover a victim's genetic defects, which may be present in other family members.

VETERINARY FORENSIC PATHOLOGY

Criminal investigations of wildlife violations and animal cruelty often involve examination by veterinarians trained in forensic pathology. They may be called upon to investigate whether an animal found by the roadside has been hit by a vehicle, has been shot, or has imbibed a toxic substance, such as radiator fluid, that has leaked from a car. Another typical case might involve determining whether an animal has been caught by a legal or illegal trap. The legality of traps varies from place to place. Leghold traps catch animals by a limb. Other kinds grab them around the body. Occasionally, the markings left on an animal's carcass by a trap may be obscured by fur. Removal of the skin allows veterinary pathologists to detect such markings and identify the kind of trap used from the distinctive pattern it leaves on an animal's body.

An animal trap, shown at left, may leave distinctive markings on the trapped animal's body, helping veterinary pathologists determine whether an animal was caught illegally.

The Autopsy

Although an autopsy may not be pretty for the average person to visualize, it is fundamental to a forensic investigation of crimes that result in death. Of course, an autopsy may be performed on someone who has died from natural causes—to determine the specific impact of a disease on the body, for example—but when a crime is involved, an autopsy is essential.

Plainly speaking, a forensic autopsy is a medical examination of the body to discover and legally record the cause of death as soon as possible after it occurs. During an autopsy, the pathologist observes the body both inside and out, taking into account all abnormalities, both gross and minute, that may have contributed to death. Like major surgery performed on a living person, it is an exacting, lengthy process.

THE AUTOPSY PROCESS

During an autopsy, which usually lasts three or four hours, the pathologist notes information that will be covered in the autopsy report: cause of death, mechanism of death, and manner of death. The cause is the specific action or condition that ended life, such as "a gunshot wound that penetrated the head" or "asphyxia." The mechanism is the change in the body that

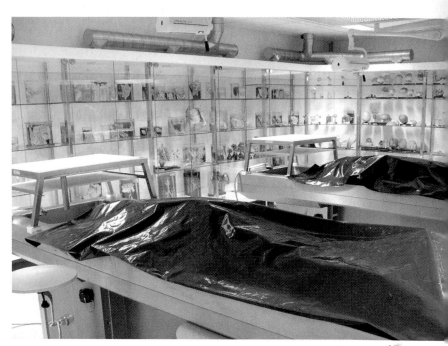

Above: Bodies covered with plastic await autopsy in a medical facility. Right: The bone saw can be used to saw through the skull. Top left: A scalpel, a common instrument used in autopsies.

makes life processes impossible, such as "constriction of the trachea to prevent intake of oxygen." The manner, as defined by law, can be natural, accidental, homicide, suicide, or, when results of pathological examination are inconclusive, undetermined. In determining the manner of death, the pathologist makes a reasoned interpretation based on evidence from the examination.

Forensic autopsies are usually conducted in a morgue. Before the autopsy actually starts, the pathologist may have X-rays taken, especially if the case involves a shooting or stabbing. The procedure begins with weighing and measuring the body and

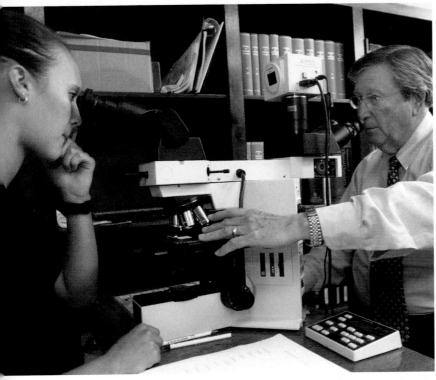

A pathologist, along with an intern, examines an autopsy tissue sample through a microscope.

noting any external marks. The pathologist is then ready to go inside. The body is opened with a major incision, in the shape of a Y, its branches extending from shoulder to shoulder and down the length of the trunk. After the skin, muscles, and other tissue have been pulled back from the sides of the incision, the rib cage is removed; the internal organs are then severed from their attachments and removed, either as a whole or individually, according to the needs of the autopsy. The skull is also opened and the top is removed like a cap, with a circular cut all the way around the cranium. The brain is extracted once it is detached from the tissue connecting it to the bone of the skull.

CLOSE INSPECTION

Throughout the procedure described above—here much simplified—the pathologist closely inspects various parts of the victim's body, sometimes drawing samples of fluids such as blood and urine for testing in the laboratory. Sections are taken from organs for laboratory examination, blood vessels are opened and scrutinized, organs are weighed, and any abnormalities found are studied.

An autopsy may seem like a bloody process, but it is not. Because the pumping of the heart stops with death, a dead body has no blood pressure. Although some blood may seep from incisions, an autopsy is by no means gory in that sense.

A victim's psychological profile, including factors such as depression, may be useful to pathologists when the manner of death seems undetermined.

PSYCHOLOGICAL AUTOPSIES

Although often unacceptable as evidence in court, reconstructions of a victim's life situation can be helpful in determining the manner in which he or she died. Known as a psychological autopsy, re-creating a picture of someone's lifestyle and frame of mind can be especially useful when the manner of death is undetermined, as it sometimes is in the case of suicides.

Investigators and mental health professionals can try to piece together a victim's psychological condition before the death occurred by interviewing family, friends, and coworkers, and even witnesses present in and around the death scene. A survey of the victim's medical history helps, as does reviewing the scene from a psychological perspective. Information such as whether the victim was depressed or worried about financial debts can help guide a pathologist to know what to look for during a physical autopsy.

Asphyxiation

When a person is blown to bits by a terrorist's bomb or a natural gas explosion or is decapitated for whatever reason, the cause of death is obvious, even to an inexperienced eye. Often, however, the way in which someone died is not readily apparent and must be deciphered by a pathologist. This can be the case when a person is asphyxiated, which occurs when the body is deprived of enough oxygen and oversupplied with enough carbon dioxide to stop it from functioning. Asphyxia—succinctly, when breathing stops—can be triggered for several reasons, including electric shock, certain poisons, diseases such as asthma and pneumonia, drowning, choking, and smothering. The latter three causes are the circumstances most often involved in homicides.

DROWNING

The classic sign of a drowning is a thick cone of pink and frothy foam covering the victim's mouth and nostrils. This material is foamy mucus, which quickly builds up in the windpipe and throat when the victim chokes from inhaling water. It is the foam that actually stops breathing. Since drowning begins with the inhalation of water, or other fluids, the stomach and lungs are filled, and sometimes contain small aquatic life as well.

STRANGULATION

When one person strangles another, either by hand or with a rope, wire, or other instrument, markings may appear on the skin of the neck and throat. Yet, even if these are lacking, a pathologist can find internal signs. Among

Above: Eighty-four passengers died of asphyxiation in a 1903 Paris Metro fire. Top left: Drowning begins with the inhalation of water or other fluid.

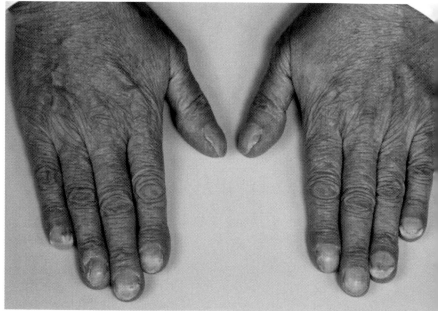

Cyanosis, or bluing of the nail beds or the lips, is evidence of inadequate oxygen in the blood. The condition is common evidence of suffocation.

An 1865 photograph shows the hanged bodies of four coconspirators in the assassination of President Abraham Lincoln. Death by hanging may result from asphyxiation, an insufficient flow of blood to the brain, or by breaking of the neck.

them are an enlarged right ventricle of the heart, unusual bulging of the veins above the place where pressure was applied, and a symptom called cyanosis, a bluish discoloration of the lips and even the fingertips. Although it would seem logical that in the case of someone who has died from hanging, the cause of death is strangulation, it is not always the case. Although hanging does constrict the passage of air, it also cuts off blood to the brain and, especially if the drop is violent, breaks the neck.

SMOTHERING

Outwardly, smothering is difficult to detect because items typically used in the process, such as a pillow or plastic bag, leave virtually no marks on a victim. Hands used to smother someone may leave external traces, but this is by no means guaranteed. In some circumstances, however, the pressure applied to a victim's mouth will leave tiny cuts and abrasions inside the lips. Usually, moreover, minute hemorrhages, which look like red pinpricks,

dot parts of the face, particularly around the eyes. Even if a pathologist singles out drowning, strangulation, or smothering as the means of death, and natural causes are ruled out, the involvement of another person may not be readily evident. This is when further pathological examination comes into play, as do the many other aspects of forensic science and the ability of investigators to piece together what happened from evidence gathered at the death scene as well as from witnesses.

Knife Wounds

To an experienced pathologist, wounds can tell a vivid
story of a homicide—not only
how a person died and the type
of weapon used, but also the
circumstances surrounding the
death, whether the victim fought
back, and where the assailant
was in relation to his target. In
a hypothetical and abbreviated
example, two men argue in a
bar and one is expelled from the
establishment. Angry, he waits in
the deserted alley next door for
the other to leave and, when he
does, a confrontation ensues. A
fight breaks out and one of the
men pulls a knife. The other
grabs for the weapon, trying to
ward it off, but the man with the
knife is quick and experienced.
He slashes, and then stabs with
the flat of the blade horizontal,
so he can rip it out more easily
for another thrust. But no further thrusts are necessary. The
victim falls to the ground, dying.
Panicked, the knife wielder
drops his weapon and runs.

MATCHING THE KNIFE
TO THE WOUND

If, in such a case, police retrieved
the knife, a pathologist would
examine the wounds in hopes
of matching it with the weapon.
The wound may be able to answer many questions. How long
was the knife? Was it single or
double edged? Did it have a hilt?
Was there a struggle?
A wound from a
single-edged knife is
usually shaped rather
like a U at the top and
a V at the bottom. That
from a double-edged knife is
V shaped at both ends. Some
knives have serrations on part
of the spine, which leave jagged
cuts. If a hilted knife completely
penetrates the body, the hilt may
leave a crosswise marking on the
skin. The width and length of
the wound, which can be clues to
the dimensions of the knife that

*Above: The body of a stabbing victim lies
bleeding on the sidewalk. A pathologist
can examine the knife wounds to help
determine what kind of weapon was
used. Below: Different shapes of
knife blades would inflict
different sorts of wounds.
A clip point is shown here.
Top left: A normal
single-edged blade.*

inflicted it, can be gauged, but contraction of tissue can sometimes obscure markings. So too can enlargement of the wound purely by the force of the knife blow and not the blade itself.

PICTURING THE ASSAULT

The nature of knife wounds can disclose the details of the attack. Slashes on the arms and hands indicate that the victim tried to fight off the attacker. Pathologists call these cuts "defense wounds." Wounds that occur when the knife is jammed into the body repeatedly while the blade is still inserted tend to be longer than those made by quick in-and-out thrusts.

Even the most competent pathologist, however, may not be able to figure out all elements of a knife attack from the wound's appearance. If the knife entered the victim's back, for example, one might surmise that the assault came from behind. If the puncture was through the chest from below, it may appear that the assailant was shorter than the victim. Yet, neither surmise may be necessarily correct. There is often a great deal of movement involved in a knife attack. Attacker and victim may grapple, bob and weave, and roll on the ground. A stab in the back may be delivered over the victim's shoulder. One that enters the victim's thorax from below may be delivered by the killer from a crouch. Again, the work of the pathologist must be coupled with that of other forensics experts and investigators to solve the crime.

A broken bottle can be used as a murder weapon. This kind of weapon inflicts a fatal wound that is identifiable to a forensics expert by its jagged perforations.

ABOUT KNIFE WOUNDS

The state of a knife wound can reveal hints about when it was inflicted. A wound that was delivered before death occurred is gaping and bleeds more heavily than one inflicted well after the victim expired—to cover up the real cause of death, for example. A knife that has been twisted or jerked about after it has entered the body leaves a much more jagged wound than one that penetrates cleanly.

Weapons other than knives can also leave puncture wounds. An ice pick, traditionally used in some mob killings, has no angular edges but instead is circular. A stab from the broken neck of a bottle can leave a mark that is somewhat circular but has jagged perforations.

This photograph of a murder victim was used as evidence in a homicide case. The shape of the stab wound in the chest, at lower right, indicates that a knife was used. Marks on the victim's neck indicate strangulation with a cord of some kind.

Bullet Wounds

Under the glare of lights in a county morgue, a pathologist begins the autopsy of a body that has been found in the woods by a hiker and retrieved by police. The pathologist, who has seen many gunshot victims before, can easily tell that the impact of a bullet created the hole in the victim's chest. The lack of a wound on the back of the body indicates that the bullet must be inside. Without even probing the hole, the physician already surmises where the shooter was when he fired: only a few yards in front of the target, but not at point-blank range.

Gunshots leave evidence that is more easily detectable by a pathologist than do many other types of wounds, such as those produced by knives. In part, this is because different types of bullets and the firearms that discharge them have their own highly distinctive signatures and because the line of fire and the distance between the gun and the target have a profound influence on the place where the bullet enters the body.

WOUND TRADEMARKS

The appearance of a wound at the point where a bullet enters the body, the "entrance wound," provides a solid indication of the distance from which the person holding the gun fired.

Pathologists break down wounds into three classes. A contact wound is from a bullet fired point-blank, with the gun's muzzle against the body. An intermediate, or close-range, wound is from a gun discharged a short distance away, generally between three and seven yards. Beyond that, the gunshot is considered long range.

Each type of bullet wound leaves a trademark appearance. A contact wound may be edged by jagged lacerations, caused by gases exploding from the discharged cartridge and blasting into the skin. Typically, the skin surrounding the bullet hole looks sooty from powder burn and is rimmed by the distinctive imprint of the muzzle where it was rammed against the body. The muzzle imprint, called an abrasion ring, and lacerations are absent from an intermediate wound, but present is what pathologists call a "tattoo" of powder, the stain of powder grains that are imbedded in the skin; this "tattoo" expands with the distance between the shooter and the target. A limited area of powder grains may also be evident in close-range wounds.

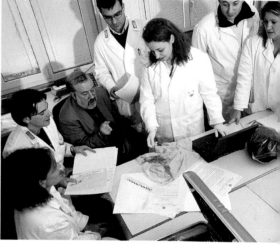

Above: Forensic scientists examine a blood-stained dress from a gunshot victim. The pattern of the blood and the bullet hole can reveal the distance from which she was shot. Top left: A gunshot releases a cloud of gas.

Shots from longer distances puncture cleanly, leaving a circular hole that may be similar in size to the caliber measurements of the bullet. The patterns described are from the best of all possible forensic worlds. In many cases, the characteristics of entrance wounds may be obscured by clothing or by decomposition of the body. Soot and powder grains may not reach a wound under heavy clothing, which can also cushion against muzzle impact. Desiccation can shrink the margins of a wound.

Most bullets remain in the body after they strike, but many are designed so that they impart

BULLET-WOUND CAVITIES

Bullets that kill quickly, especially on impact, are said to have great shocking power. High-velocity bullets generally have the most shocking power, and their effect on the body at the instant of impact sends a pressure wave through the surrounding soft tissue, causing it to balloon outward. The shock wave that moves through the body might be likened to the splash of water that erupts when a rock falls into a pond. The ballooning creates a cavity in the tissue, stressing and damaging tissue far from the actual point of impact. This cavity is temporary and soon subsides. Another type of cavity is formed as the bullet smashes along its direct path, destroying tissue and causing direct damage as it goes. Traditionally, pathologists and ballistics experts gave velocity almost all the credit for forming large temporary cavities, but research has shown that even at lower velocities, bullets that deform with impact do the same.

High-velocity bullets have the most shocking power, but even bullets that travel at lower velocities can cause significant damage to a human body.

the maximum amount of kinetic energy upon impact, so expend their penetrating power inside the body. Occasionally, a bullet from a high-powered round will pass through, leaving an exit wound that is generally somewhat larger than the entrance wound. Large exit wounds are more likely to occur when a bullet does not maintain a straight trajectory but instead tumbles within the body. Large exit wounds also result from bullets designed to cause greater tissue damage, such as hollow points, which fragment, and soft-points, which spread out before they plow through the victim.

If both entrance and exit wounds are present, investigators will attempt to recover the bullet by extending the path it took through the body. To recover bullets lodged internally, pathologists often use long metal probes as well as X-rays, fluoroscopy, and sometimes metal detectors. Once a bullet is found, it is delivered to ballistics experts, who attempt to match it to the gun that fired it.

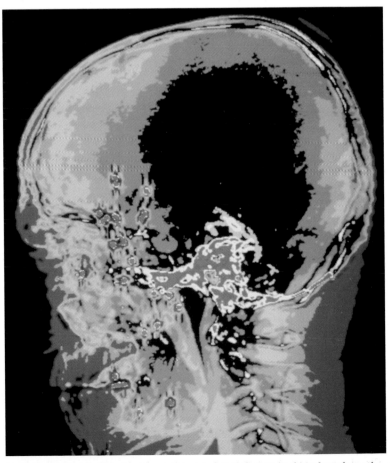

Lead shots in the skull and brain are shown as hot pink spots in this color-enhanced CT scan. This gunshot injury was the result of a hunting accident.

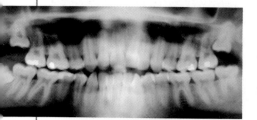

A Different Kind of Dentist

Even a few teeth can provide compelling clues to their owner's identity. Deciphering those clues is the goal of forensic dentistry, more precisely known as forensic odontology. Equally important is the role of the forensic dentist in using human dental remains to identify unknown victims. This is particularly the case after catastrophic disasters, such as earthquakes, bombings, and airplane crashes.

IDENTIFYING THE UNKNOWNS
Forensic dentists note the features of teeth and other parts of dentition, visually and by X-rays similar to those normally performed in any dentist's office. If jaw and skull remains are available, these may be included in the examination. Once dental characteristics are defined, a search begins to match them with existing records in police files or the files of dental practitioners or hospitals. The use of computerized records has made the search easier, but it remains a

monumental task. Nevertheless, it is an effective process, bringing some form of closure to the families of the deceased.

Forensic dentists stress that police missing persons reports should include substantial information about the individual's dental history. Police should ask the person filing the report for the names of the missing individual's family dentist and specialists; whether he or she served in the military; and whether regular dental visits occurred. To gather as complete a picture of a missing person's dental history as possible, forensic dentists often ask for information from health insurers, public aid agencies, and even coworkers.

The techniques used in recovering dental evidence from a body may vary according to the condition in which the body was found. The teeth of decomposed bodies are usually intact, but breakdown of the gum fibers that link teeth to the bone may have permitted teeth to fall out. They may be strewn on the

ground or even in the back of the mouth. In the case of burned bodies, it is essential that teeth are handled with extreme care because heat may have so desiccated and charred them that they may shatter at a touch.

BITE MARKS
Humans probably have been using teeth as weapons since appearing on this planet. Unlike fingerprints, human dentition patterns have not been scientifically proven to be unique, although solid evidence exists that the patterns left by the biting edges of the upper and lower front teeth are unique to each person. The marks left by human teeth on a victim, however, can at least point in the direction of the person who inflicted them. Biting can be an element in several types of crime. During a violent physical confrontation, the aggressor, the victim, or both may bite an opponent. Rapists and perpetrators of sadistic crimes often inflict bites on their victims during their assaults. Serial killer Ted Bundy was notorious for

Left: Teeth with fillings are especially useful in identifying corpses. Teeth from decomposed bodies are usually intact, though they often fall out; they can be matched with missing persons' dental records. Top left: Forensic dentists use X-rays to gain evidence about both crime victims and perpetrators.

During the murder trial of serial killer Ted Bundy, forensic dentist Dr. Richard Souviron points to a photograph of Bundy's teeth, demonstrating to the jury that the dental patterns shown matched bite marks found on one of his victim's bodies.

ARE BITE-MARK PROFILES POSSIBLE?

In the early 1950s, the members of a teenaged gang in a large Midwestern city were notorious for biting rival gang members during brawls. Their reputation was such that they were known as the Fang Gang. The Fang Gang has gone the way of boppers and saddle shoes, but biting during violent confrontations occurs far more often than most people realize. By prevailing standards of medicolegal science, however, it is not yet possible—and it may never be—to create bite-mark profiles with the same accuracy with which genetic profiling of an individual's DNA is accomplished. The features of an individual's teeth and bite can be matched, but it is yet undetermined if these traits are unique, at least as far as the courts are concerned. For example, it stands to reason that more than one individual can have a missing right canine tooth. The lack of a particular tooth, on the other hand, can help investigators focus on one suspect and eliminate others.

biting his victims on several parts of the body, an action viewed as an attempt to demonstrate dominance. Children who are victims of child abuse sometimes suffer repeated biting by their tormentors. Whatever the reason, biting usually signifies an event of vicious, unbridled violence.

Increasingly, pediatricians, dentists, school nurses, and other practitioners who come in regular contact with children are urged to be alert for indications of bite marks and, if their presence is suspected, to contact appropriate authorities. The wounds vary according to the type of teeth. Canine teeth create triangular marks, while those left by incisors are rectangular. The condition of a person's teeth, caused by wear, breakage, dental appliances, and disease, may alter these markings but also can serve as identifying traits.

Anyone who has sat in a dentist's chair is familiar with some of the methods used by

Dental impressions such as this cast can be made of a crime suspect's teeth, allowing forensic dentists to compare the position of teeth to bite wounds found on a victim's body.

forensic dentists. X-rays and casts of teeth impressions from suspects are typical tools in forensic dentistry, and photographs of bite marks are used to record clues. Although lacking the legal power of fingerprints, the condition, shape, and position of teeth in a bite mark can be powerful evidence in tracking, and sometimes convicting, criminals.

CHAPTER 5

FORENSIC ANTHROPOLOGY

Left: Forensic scientists use facial reconstruction to re-create the likeness of a crime or disaster victim as an aid to identification. A plaster cast of the skull is covered with clay modeled around pegs that indicate the depth of facial muscles. Top: The documentation of a crime scene may be time-consuming, but it is a critical step in processing it accurately. It is the last chance to preserve a permanent record of the condition of the scene and any physical evidence within it. Bottom: Water environments present investigators with difficult challenges, such as when remains drift and become disarticulated, leaving parts behind.

According to the Bible, in a vision, the prophet Ezekiel heard God ask, "Can these dry bones live?" The answer was affirmative and "tendons and flesh appeared on them and skin covered them." Today's forensic anthropologists may not be able to literally awaken life in dead bones, but they can often determine to whom they belonged, and when their owner died. As for putting flesh upon them, with the help of forensic artists, a close approximation of a person's face can be overlaid on a skull, not with living tissue, but with plastic, plaster, clay, or digitalized features.

Forensic anthropology is a relatively new field in the application of science to law enforcement, gaining prominence long after pathology, toxicology, and other scientific fields were routinely used to solve crimes. Examination of bones and measurements of human body parts were employed to a degree in investigating homicides during the nineteenth century, but not until the middle of the twentieth century was the law enforcement establishment fully aware of anthropology's potential as a forensic technique. Once established, it became one of the most important and widely used disciplines employed in the forensic sciences.

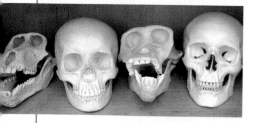

The Role of Forensic Anthropology

The anthropologist at work, in the public's mind, is often pictured as someone sifting the bones of ancient hominids from million-year-old sands. This perception is accurate, to a degree, but the same techniques and processes that are used to trace the evolution of the human species also are utilized by anthropologists who work in forensics.

By definition, anthropology is the science that studies humans, their origins, and their evolution, both biologically and culturally. Facets of human culture, such as coming-of-age rites among South Sea islanders, constitute one branch of anthropology. It is the other branch, physical anthropology, that deals with humanity's evolutionary and physical position among the order of primates, which has aspects that are applied most frequently to the legal system.

INVESTIGATING BONES

Basically, the physical anthropologist investigates bones: both old, especially very old, bones in the study of humans and human evolution, and those that are recent, often relatively fresh when the purpose of investigation is forensic. In some ways, the tasks of the forensic anthropologist parallel those of the pathologist, in that the role of both scientists

Above: Human bone tissue can reveal information about a victim's identity and cause of death long after the soft tissues have decomposed. Top left: Modern investigative techniques employ many of the same methods that anthropologists use to understand human evolution.

is to examine human remains. Pathologists concentrate more on the body's soft tissues and need a body, or at least body parts, to perform an autopsy. Forensic anthropologists work largely with the bones that remain after the body has decomposed. They often work on cases that have gone

unsolved for prolonged periods of time and are thus surrounded by mystery; these cases often capture headlines as well as the public's imagination. Either way, the contributions of pathologists and anthropologists help a coroner or medical examiner estimate the cause and manner of death.

Forensic anthropologists need expertise in the structure, development, and composition of bone, as well as in skeletal biology and anatomy, and often have a working knowledge of fields such as genetics, human development, archaeological excavation, and even entomology, because insects sometimes impact bones during decomposition.

By examining bones, the forensic anthropologist seeks to help reconstruct a profile of the individual to which they belonged and whose other remains are either gone or unrecognizable. This profile includes, ideally, the identity, but also the sex, ancestry, age at death, stature and body dimensions, injuries and disease, time elapsed since death, whether death was violent, what happened to the body after death, and sometimes even the individual's vocation, diet, and place of origin. From time to time, especially when bones are fragmentary, the anthropologist must first determine whether they belong

Physical evidence found at a site, such as these eighteenth-century musket balls, provides clues that help forensic anthropologists establish a context for a crime scene.

clean bones for examination, and also work in tandem with police, pathologists, odontologists, and other forensic specialists. Like an archaeologist, moreover, a forensic anthropologist needs an eye for artifacts present at the crime scene or otherwise associated with a body. We can construct a hypothetical case in point: Perhaps a skeleton is discovered in the basement of a colonial-era building that is being demolished. A hole, perhaps from a projectile, has punctured a piece of the skull. Given the age of the building and the appearance of the bones, an experienced anthropologist might suspect that the victim was dispatched long ago. If, among the bones, he found a ball from an eighteenth-century musket, he could surmise that his suspicion was correct.

to human or animal, a process that is not as easy as it might seem and often requires microscopic examination of cross sections, X-rays, and chemical testing.

HANDS-ON SCIENCE

Forensic anthropologists are often on the staff of institutions such as museums and universities and work with law enforcement agencies on a consulting or on-call basis. Although they are not full-time scientific detectives, forensic anthropologists are very involved in hands-on tasks. They may work at the crime scene unearthing and retrieving human remains,

EVOLUTION OF FORENSIC ANTHROPOLOGY

The partnership of forensic anthropology and law enforcement began to blossom in the 1930s, when the FBI began engaging Smithsonian Institution anthropologists to assist in investigations, a working association that continues to this day. FBI director J. Edgar Hoover realized that the agency's fledgling crime laboratory had no expertise in anthropological science and fostered the collaboration between the two federal entities. Aleš Hrdlička (1869–1943) was the Smithsonian's first curator of physical anthropology, and collaborated with the FBI on several cases, therefore establishing the foundations of forensic anthropology. Other Smithsonian anthropologists have followed in his footsteps. Another pioneer of forensic anthropology was Harry L. Shapiro (1902–1990), who was on the staff of the American Museum of Natural History and Columbia University in New York City. After World War II, he was asked to help identify the remains of unknown war dead in France. Shapiro applied established techniques of physical anthropology to the task and formulated methods for identifying victims among large numbers of remains. Later, Shapiro's expertise was in great demand by law enforcement agencies and foreshadowed the importance of forensic anthropology in today's legal system.

The American Museum of Natural History served as one of the pioneering institutions that brought the scientific techniques of forensic anthropology to the practice of law enforcement.

The Study of Bones

It is only natural that because forensic anthropologists are concerned with bones, the scientific study of bones, called osteology, lies at the heart of their experience. The ability to aid in the identification and cause of death of skeletal remains, as well as the recovery of remains using archaeological techniques, involves detailed knowledge of skeletal anatomy and biology. The discipline of osteology covers just about everything associated with bones, from how they form and grow to characteristics such as hardness and their appearance before and after death.

GAUGING AGE AND SIZE

The bones that support and protect the soft parts of the human body may seem inert, but, in reality, they are living tissue, just as much as flesh and blood are alive. Bone—technically called osseous tissue and formed of calcium compounds—is built of cells, osteoblasts and osteoclasts, that, as do other living cells, depend on a blood supply to function.

Bones in the human body grow in size from birth to late adolescence, often, as parents of teenagers know, in dramatic spurts. Although bone growth eventually does cease with age, damage control continues throughout a person's life. When bone is injured, cells mobilize quickly to carry out repairs. If you break a finger, for example, blood quickly clots at the site of the break; then, cells start manufacturing new bone in order to bridge the fracture. The scar that remains from the break is unique to each individual's injury and can serve as a marker by which a forensic anthropologist can identify remains, by comparing the characteristics particular to the bones with X-rays of any likely candidates. Since the bones that make up the average skeleton are very hard, they would be the parts most likely left of a corpse.

Although bones grow in size as a person matures, their number actually decreases. A newborn baby has more than 300 developing bones in its skeleton; each parent has only 206. The reason for the drop in number is that certain bones fuse together. For example, in the long bones of the infant the long tubular part, or diaphysis,

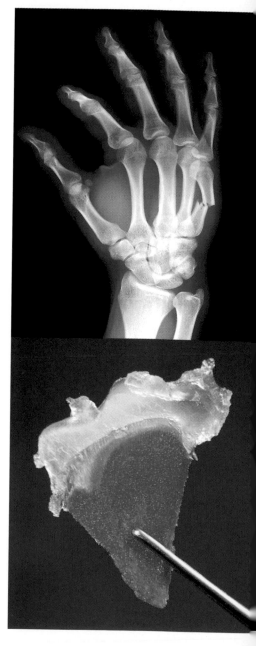

Top right: When a broken bone heals, it leaves permanent scars that, when compared against a medical history, can be used to identify a victim. Bottom right: Examination of cartilage from the growth plate border of a bone may indicate a growth abnormality. Top left: Bones can hold clues to the fates suffered by their owners long after other evidence has disappeared. This ancient skeleton belonged to a battle victim who suffered a fatal wound caused by a sword.

is separated from the bony caps on the ends, or epiphyses, by cartilage, allowing for growth. At various stages of maturity, these components fuse to form single bones and longitudinal growth ceases. Measurements of the length of these mature bones, especially those in the legs, can be used to calculate living body stature, or height.

The growing bones of the infant skull are separated by strong, fibrous elastic tissue, in some areas known as fontanels, a baby's soft spot. Even after the soft tissue in the fontanels is replaced by bone, the individual bones remain separate until maturity, their lines of separation referred to as sutures. During the adult years, even the sutures may disappear, resulting in the union of adjacent cranial bones. The extent of adult cranial suture closure increases with age but is highly variable.

DIFFERENCES BETWEEN THE SEXES

Humans in general exhibit a low level of sexual dimorphism, or the systematic difference in form between individuals of different sex in the same species. Nonetheless, there are pronounced sexual differences visible in a few bones. While the average male body has slightly thicker and longer appendages, other factors such as nutrition can affect this statistic.

A staple of sexual identification for the forensic anthropologist is the pelvis, the bony ring of hip bones that supports the weight of the torso. Because it must permit childbearing, the female pelvis is

relatively wider than that of the male. There are also sexual differences in the comparative lengths of specific bones in the pelvis, notably the pubis, or lower front, and the ischium, or lower back, of the pelvis. Other bones that may display sexual differences are the rib cage, which tends to be narrower in females, and the jaw, which tends to be more prominent in males.

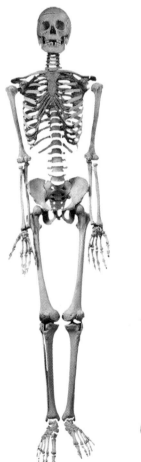

Above: Skeletal characteristics can help to determine body size or approximate ethnicity. Right: The sex of a victim is most easily determined from the pelvis, which in females is relatively wider than in males.

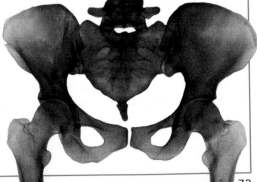

Many of the tools used by forensic anthropologists have remained unchanged for generations. This spreading caliper, for example, is used to measure the diameter or thickness of an object, such as a skull.

STILL IN STYLE

Many of the instruments that forensic anthropologists employ to measure bones are not in the realm of the high-tech but tools that have been in use by generation upon generation of scientists. Among the most frequently utilized is the traditional caliper. There are different styles of caliper for specific tasks. Spreading calipers, with gooseneck arms, are designed to measure the space between points on the skull. The sliding caliper, which resembles a slide rule, measures bones of the face and jaw. Another device, the osteometric board, measures long bones, such as the tibia and femur, the latter of which is the longest bone in the human body.

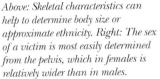

Reading Bones

A pile of bones found in the basement of an old building or in a shallow grave may look like a gruesome, jumbled mess to the average person, but to a forensic anthropologist it represents reading material of a sort, telling the story of the fate met by the person to whom the bones belonged. Competent forensic anthropologists can read bones like a book, gathering testimony that, while it may be mute, can be powerful evidence in a court of law. More often than not, bones are treated as evidence of last resort, since by the time they are found, other clues may have long disappeared.

INTERPRETING THE SKULL

By virtue of obvious gaps between skull plates and gross differences in the pelvis, age and sex are traits that generally are the easiest to identify from bones. Female eye sockets have sharper margins than those of males and the bony ridges above the eyes are not as prominent. Less obvious, and by no means always

accurate, are distinctions between ancestries. As a rule of thumb, however, the crania of individuals of European descent tend to have nasal openings that are comparatively narrow, with narrow faces

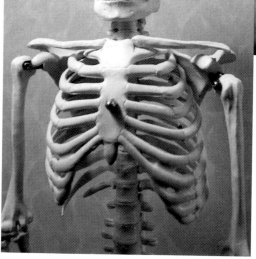

and prominent chins, while the nasal openings of the crania of individuals of African descent are wider. The cheekbones of Asians and Native Americans are prominent and project beyond the eye to a greater degree than those of the other ancestries.

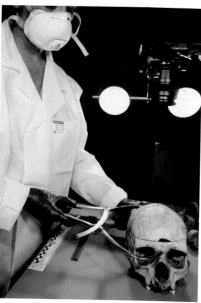

Above and left: Age and sex are usually the easiest characteristics for an anthropologist to estimate. Age can be determined by the degree of fusing in certain skull and limb bones, and the male skull is usually more muscle marked than that of the female. Top left: A skilled forensic anthropologist can glean volumes of information from skeletal remains.

BODY SIZE INDICATORS

The places where muscles attach to bones can serve as indicators of body size. Repeated expansion and contraction of muscles at the point of attachment during life build up bony ridges that can be clearly seen by the trained eye. Generally, the larger the muscles and the more they are used, the bigger the ridges, indicating whether an individual's physique is robust or slight. These bony

Left: The amount of time that has passed since a person died can be approximated from the condition of the remains. Bones that are cracked from freezing or modified by scavenging animals can indicate that considerable time has passed since death. Right: Bones can preserve details of how a victim died. The skull of this individual reveals that he apparently received a heavy blow to the head with a blunt object.

ridges also may suggest, although not at all definitively, the kind of occupation in which the individual may have been involved. This is not to suggest that one could distinguish between an astrophysicist and a molecular biologist; the ridges likely will be larger on someone who works with his or her hands, such as a carpenter, than on someone who does not, such as a librarian.

THE CONDITION OF BONES

The condition of bones can also be a gauge to the time that has expired since death. Bones with no hair, ligaments, or bits of flesh attached are probably those of an individual who has been deceased for quite some time. Bones that are badly weathered, perhaps cracked by freezing, or gnawed by animals signify the passage of considerable time since their owner expired.

There are many variables, however, such as weather, that affect the condition of bones, and careful analysis must be made

before the age of bones can be estimated with any accuracy. Like soft parts of the body, bones also may bear the marks of wounds, bullet holes, or skull fractures caused by a blow from a heavy object. Even a small piece of bone can tell of violent death. The hyoid, for example, is a

small U-shaped bone between the tongue and the larynx that is the key to the ability to speak. The pressure of hands on the throat when one person is throttled to death by another may fracture the hyoid, a dramatic indication of the way in which the victim was killed.

Smithsonian's Department of Anthropology houses an extensive skeleton collection.

The Department of Anthropology at the Smithsonian Institution's National Museum of Natural History is home to one of the world's largest and best-documented collections of human skeletons. The Robert J. Terry Anatomical Skeletal Collection has more than 1,700 skeletons of known age, ethnic origin, cause of death, and pathology. The records of each skeleton note even minute details contained in individual morgue files, anthropometric measurements, dental charts, bone inventories, and autopsy reports. Each skeleton has its own photographs and hair samples, and some of them also have a corresponding plaster death mask.

A Classic Case

One notable homicide investigation often cited as a classic of nascent forensic science at work occurred in Scotland in 1935. This case shows how various disciplines in the forensic sciences can bring even the most cunning of criminals to justice.

THE "JIGSAW MURDERS"

It all began when two young women walking along the road from Edinburgh to Carlisle looked over a bridge and saw a human arm in a stream that flowed at the bottom of a ravine below the structure. Police were called to the scene and began to search the brushy little valley, now known to locals as "Ruxton's dump." They eventually turned up 43 decomposed chunks of flesh and bone that obviously belonged to more than one body. It was also clear to investigators that the fingertips of the hands had been removed with clinical skill. When the chunks were pieced together, they composed two corpses so badly decomposed that identification would prove to be a formidable task.

"BONE FORENSICS"

Experts were called in from the University of Glasgow and the University of Edinburgh, citadels

of forensic science. Because the bodies had been in the ravine for a considerable amount of time and because the skull wears less than other parts of the skeleton, investigators focused their attention on the victims' skulls. By examining general features such as whether there was a high level of muscle marking that would indicate a male, and the eye sockets, which have sharper margins in women, the experts were able to declare with some certainty that the remains were female.

The remains turned out to be those of Isabella Ruxton, wife of prominent physician Buck Ruxton, and her children's nursemaid, Mary Rogerson. Mrs. Ruxton was positively identified

Left: In 1935, Isabella Ruxton, along with her children's nursemaid, was murdered by her husband. Below: Dr. Buck Ruxton, an Indian-born Scottish physician, was ultimately convicted and executed for murdering his wife and Mary Rogerson. Investigators created a sensation by proving their case using techniques that were essentially unheard of at the time. Above left: The University of Glasgow in Scotland was the scene in 1935 of the cutting-edge use of forensic science to solve a double murder known as the Jigsaw Murders.

by a technique that was then cutting-edge: Investigators superimposed a photo-transparency of one of the recovered skulls on her portrait, and the features matched.

TIMING THE MURDERS

Isabella Ruxton was last seen alive about two weeks before the bodies, or, more precisely, the pieces, were discovered in the ravine. Blowfly maggots hatched from eggs laid on the bodies and found feeding on the remains were calculated to be between 7 and 14 days old. Therefore, the bodies had been there for at least that amount of time.

The trail ultimately led to Dr. Ruxton, who had left plenty of other evidence behind, including the precision with

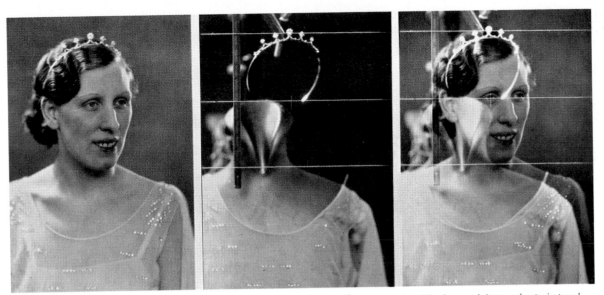

To find the precise scale of this portrait of Isabella Ruxton, a photographer staged a measured shot of the dress and tiara and superimposed them on it. The results of this "Finding the Scale" reconstruction supported the interpretation that remains found in the ravine belonged to her.

which the bodies were dismembered. Even more damning were the facts that one of the heads had been wrapped in clothing belonging to one of the Ruxton children and that a cleaning lady had seen bloodstains in the home. This physical evidence, added to reports of Ruxton's violent mood swings and threats to kill his wife because he was convinced that she was cheating on him, made a verdict in the case easy to decide. A jury convicted Ruxton of double murder and he went to the gallows.

The Ruxton case, as are so many other sensational murders solved by forensic science, was splashed across the headlines in newspapers all over the world. Ever drawn to the macabre, people lined up to buy items from the Ruxton home to serve as mementos. The investigation was a textbook example of the role that the sciences can play in forensics. The case is even more meaningful because it occurred at a time when crime laboratories were few and organized forensic science was in its infancy. The success of the "bone forensics" method used in the Jigsaw Murders case, so widely reported in the press, led to an increased public trust in forensic science, and a growing measure of professional recognition.

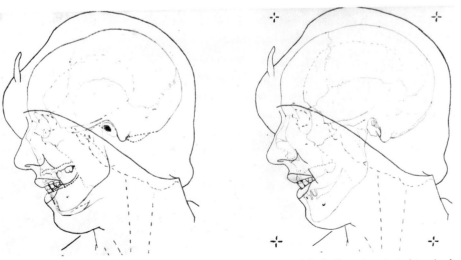

Superimposed outlines of Isabella Ruxton's face and both of the recovered skulls. Forensics experts determined by careful comparison that the facial outline of Skull No. 2, at right, corresponded with Mrs. Ruxton.

The Body Farm

Neighbors complain about this Tennessee farm's bad odors. Their jibes, however, are tongue in cheek, and, after all, some of the scents that the breeze wafts from many farms are in fact not particularly pleasing to the nose. This one, however, is not your run-of-the-mill farm, nor is the odor that rises into the air from its fields on hot, muggy days your typical smell. What is planted in the ground there does not grow, nor is it intended to: The crop that is put into this ground consists of human corpses. Welcome to the University of Tennessee Forensic Anthropology Facility, otherwise known as the "Body Farm."

A UNIQUE CROP

Started in 1972, the three-acre site, surrounded by chain-link fence topped with razor-edged concertina wire, is an outdoor forensics research laboratory designed to foster knowledge of what happens to a body once life expires. More than 300 bodies have been given to the facility, some of them unclaimed corpses from the offices of medical examiners and coroners, others are left by donors. About 50 bodies arrive annually and, as an increasing number of people learn about the farm, the list of individuals who wish to donate their remains after death is growing. Several news outlets have profiled the facility,

which is also the setting of a novel, *The Body Farm,* by mystery writer Patricia Cornwell, and the subject of a book by its founder, anthropologist William M. Bass, called *Death's Acre.*

Admittedly, the landscape of the Body Farm is rather grisly. Strewn about are corpses in a variety of situations and conditions simulating those of death scenes likely to be encountered by police and forensic investigators. Crammed into cars, stuffed into crannies, hidden in buildings, submerged underwater, lying in meadows or under brush, or buried in shallow graves, the bodies await scrutiny by students, forensic professionals, and law

Left: Dr. William M. Bass founded the Body Farm and the University of Tennessee's collection of human skeletons. Here Dr. Bass holds a human skull with a bullet wound, one of more than 300 specimens from the collection. Right: Since 1972, University of Tennessee anthropologists and students have used the Body Farm as a controlled environment to study the changes that occur to the human body after death. Top left: Unlike the typical farm shown here, the Body Farm is an outdoor laboratory where donated human remains are subjected to the elements.

A colored scanning electron micrograph of a blowfly (Lucilia sp.) laying her eggs. Blowflies lay their eggs on dead bodies, which they can detect in a matter of minutes. Forensic entomologists study the larvae, or maggots, as a way of determining how long a victim has been dead.

enforcement personnel. The object is both research and training. FBI agents, for example, regularly visit the Body Farm to dig for bodies that farm workers have prepared to simulate crime scenes in order to sharpen their crime-scene recovery skills.

Research at the facility focuses on decomposition of the human body, such as how fast it decays under different conditions and the processes involved, such as protein degradation, amino-acid breakdown, and levels of gas in the tissue. Researchers study both how the natural processes of the body itself impact its rate of decomposition and how external agents, such as weather or insect activity, affect it. Both of these factors play a role in estimating how long a body has been in a particular location, so investigators can calculate the time of death.

INSECT AIDES

Perhaps no other insect has aided more in determining the time of death than the blowfly or, more precisely, the maggots that hatch from the eggs that blowflies lay on dead bodies. Knowing the time frame for the development of blowfly eggs and maggots, investigators can infer how long they have been on the body. The time at which adult flies lay their eggs and the course of the rest of the insect's cycle are, however, influenced by environmental conditions such as climate, weather, and location, one of the reasons why bodies at the farm are subjected to such varied conditions.

These factors can also influence the internal processes by which the body decomposes, such as the breakdown of cell membranes and the resultant

seepage of the cytoplasm that cells contain. Bacteria break down cellular contents, releasing hydrogen sulfide gas—the source of the foul odor of decay—and digestive enzymes turn on the body itself for nourishment.

So many forces influence decomposition that even the most careful calculations estimating the time of death may be off target. Information gleaned from those bodies scattered about the Tennessee countryside are helping refine those techniques. And the success of William M. Bass's groundbreaking farm may spawn similar research centers in other parts of the United States. For example, in 2005, a biological anthropology professor at the University of Northern Iowa proposed a similar facility in the Midwest.

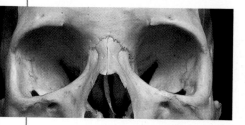

Re-creating Faces

The human skull, with its gaping, empty eye sockets and macabre grin, holds many mysteries, perhaps the most haunting of which is what the person represented by that construction of inert bone looked like when he or she was alive. Working with anthropologists and using a knowledge of anatomy as well as artistic skills, forensic artists and sculptors reproduce the facial features of a deceased person in two or three dimensions, often with uncanny accuracy. A vanished face can be brought to life by means of a composite drawing, a sculpted model, and, with digital technology, a three-dimensional computer model. Facial recreations help identify remains and, in the case of composite drawings, find suspects wanted by criminal investigators.

FROM THE INSIDE OUT

Just as a portraitist uses a model, the forensic artist does not rely on imagination when building a three-dimensional reproduction of a face and head. The model, by and large, is composed of facts and figures developed from a thorough anthropological analysis of skull features, such as the size and shape of the supraorbital ridges, or ridges above the eyes, and the appearance of the nose as suggested by remaining nasal bones. When possible, the artist also tries to visualize how other hypotheses about the suspect (ancestry, gender, body size, lifestyle, geographic origins, occupation) might influence facial appearance. All of this information is factored into the estimation of the depth of facial tissue that once overlaid the bone and used to place markers that aid in the facial reconstruction.

The skull is placed on a special workable stand that allows the artist to tilt and turn it. Artificial eyes are then inserted into the sockets, and their angle adjusted. Methods used to sculpt the reproduction generally depend on the artist's preference but usually include clay and sometimes plaster. Whichever medium is chosen, it is systematically applied directly onto the

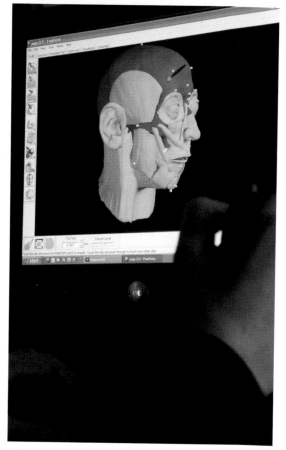

Above: Three-dimensional computer modeling enables forensic scientists to rebuild a face by digitally layering muscle tissue onto a scanned image of a skull. Top left: Reconstructing what a person's face looked like using facial bones as a framework on which to build can help investigators to identify unknown remains.

skull, painstakingly following the skull's contours and paying strict attention to the applied tissue markers that indicate angles and contours. Artificial hair, often a wig, is placed on the cranium to complete the picture. Enhancements such as a hat, eyeglasses, and other articles of clothing are sometimes added, especially if the presence of these items is indicated by evidence found at the crime scene. When the reconstruction is complete, the sculpture is photographed, and all procedures are documented and working notes collected. Well-executed facial re-creations that present a sufficient likeness of the living individual have proven invaluable in identifying unknown victims of crimes when there are no other means available.

Since the late 1980s, computers have also been used to generate facial images by digitized image capture, graphical modeling, and animation. A typical technique involves the use of a camera to digitalize the features of the skull. Sophisticated computer software is then used to adjust the features to agree with the conception of the face derived from anthropological assessments.

AGE PROGRESSION

Computers are also used to hypothesize what the appearance might be of a person missing for a very long time, or of a criminal wanted for an extended period. Before computers were involved, this technique was done by hand; in some cases, it still is. The technique, known as age progression, relies on both scientific knowledge of how aging affects facial features and the artist's experience and intuition of how variables such as hair loss and weight gain sculpt a person's appearance with time.

COMPOSITE DRAWINGS

Although outside the realm of forensic anthropology, composite drawing is the oldest of the techniques used to re-create faces and, even in the age of digitalized images, remains an important tool and the most widely used for identifying missing people and criminal suspects. Drawings are especially useful because they can be disseminated through the media, fliers, and posters, and to home computers via e-mail. As with other techniques, the artist who creates a composite drawing relies on whatever human remains, such as a skull, are available. Frequently, however, no physical material is available and the artist relies on descriptions from witnesses to complete the sketch. Time and time again, experienced forensic artists have reproduced the appearance of the missing and the wanted with remarkable accuracy. The advantage of sketching is that only standard artist's tools are required and the resultant product can be distributed inexpensively.

The process depends on access to original material such as old photographs or sketches of the individual. Specialized graphic software is not needed for this technique because artists can use standard commercially available programs to create changes in features that are estimated to occur over the years.

Artificial eyes and hair are used in facial reconstruction to complete the picture and create a genuinely lifelike appearance.

Human Rights

Anthropologists are leading a profound new trend in the forensic sciences, one with monumental global importance. It is a task that, while not new, has taken on critical importance in an age when unconventional regional conflicts, such as insurgencies and civil wars, erupt across the world, and thousands of individuals massacred by dictatorial governments or rival factions lie under the earth in places their murderers hope will never be discovered. The mission: documentation of crimes against humanity.

IDENTIFYING VICTIMS

Spurred on by groups such as the American Association for the Advancement of Science, which has a human rights committee, forensic anthropologists have exhumed mass graves in places such as Iraq, Bosnia, Guatemala, and Rwanda and examined the contents therein. The objective of these forensic investigators is the same as in the case of individual homicides: to identify the victims and lead law enforcement to the perpetrators, but on a scale seldom before attempted.

In the Balkans, forensic anthropologists working under the International Criminal Tribunal for the Former Yugoslavia, established by the

Above: English and Bosnian forensic experts inspect body remains at a mass-grave site near Sarajevo, Bosnia, in 2005. Forensic investigation of a mass grave is similar to that of a crime scene of an individual homicide, but on a much larger and more challenging scale. Top left: Around the world, human rights are violated with alarming frequency.

United Nations, have exhumed and examined thousands of victims to document war crimes committed during the conflict there. Two thousand bodies were exhumed and autopsied in Kosovo during 2000 and 2001. During the 1990s, forensic anthropology teams probed graves in Guatemala, where in

the 1980s, the nation's military carried out a counterinsurgency effort that some charged was an extermination campaign against rural Mayans. Many of the victims were killed without regard to age or sex by paramilitary groups sponsored by the government. Under the auspices of the United Nations, anthropologists documented genocidal killings during the 1990s' civil war there. More recently, teams exhumed mass graves of Iraqis allegedly slaughtered by Saddam Hussein before his fall from power.

WAR CRIMES

Forensic anthropologists working on war crimes have objectives and priorities that are somewhat different from those of their colleagues engaged in more typical cases. Their role impinges on that of a pathologist and even a criminal investigator, because they are seeking to determine not only the time since death and identification of victims but also the cause of death. They look for and examine not only remains but also other evidence, such as shell casings, possessions of the deceased, and marks from earth-moving equipment used to cover the graves. They also carry out the heartbreaking job of interviewing survivors and witnesses, a function that often demands knowledge of local cultural attitudes to be effective and requires extensive sympathy and commiseration.

When an individual person is reported missing, forensic anthropologists have at the least an idea of whom the bones they examine might represent. This is not the rule when the number of missing people, who often are unreported because kin are dead as well, is in the unidentified thousands—250,000 disappeared in Bosnia alone. To make matters worse, the bones of individuals in a mass grave are often mixed together

Land mines left scattered throughout the countryside are just one of the many threats that forensic anthropologists face when conducting investigations in war-torn areas.

in a ghastly jumble. The task of figuring out even how many people the remains represent is daunting. Dental and medical records are seldom available, if they ever existed at all.

To prove war crimes, moreover, the remains must be sorted out demographically, by age and sex, so men of fighting age can be distinguished from the elderly, women, and children.

Demographics can counter arguments from the accused that the fatalities were the result of accepted forms of combat and that the victims were military combatants or insurgents. Added to these problems are the physical dangers not encountered in an anthropological laboratory, not the least of which are land mines remaining from the time of conflict.

Machine-gun bullet casings, such as these .50 caliber, may signal a human rights crime.

CHAPTER 6

TRADITIONAL AND TESTED

Left: Forensic scientists use a gas chromatograph (CG) to find out how many components are in a mixture by analyzing the chemical composition of the gaseous and volatile liquids that result over a period of time and a range of temperatures. Top: Although fingerprinting has been used by law enforcement since the nineteenth century, electronic databases have resulted in faster matching. Bottom: Laboratory techniques reproduce possible real-life scenarios to prove or disprove suppositions in an investigation.

Never before has forensics seemed so much at the cutting edge of science. As researchers explore the human genome and the intricacies of DNA, forensic scientists track criminals by clues garnered from the criminals' own beings and that of their victims. Devices and techniques with tongue-twisting names—gas chromatograph, mass spectroscopes, tetramethylbenzidine, and the like—are routinely brought into play to help solve crimes. In danger of being lost amid the high-tech glamour of super science are traditional techniques and disciplines that have been part of organized forensics since its beginnings and remain at its core today. For example, law enforcement still relies on tried-and-true fingerprinting as a means of forensic identification. Fingerprinting has even gained increased effectiveness by coupling it with advances such as electronic data banks.

Other traditional techniques are also enhanced by access to the mass matching and retrieval systems of national and international data banks. Police can quickly compare evidence, such as markings on spent bullets. Examiners of questioned documents still use hallowed techniques such as handwriting analysis but are also adept at ferreting out incriminating hidden computer files.

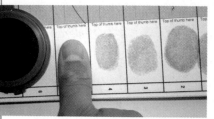

About Fingerprints

Even before you were born, you possessed the fingerprints that are yours and, according to most experts, yours alone. Fingerprints begin forming on the skin of the digits when a fetus is five months old, are clearly defined at seven months, and remain there for life. If the skin is damaged, the fingerprints grow again, although significant scarring may interfere with their reappearance. Visible as fine lines that form looped, whorled, or arched ridges on the epidermis (the thin, outer surface of the skin), fingerprints arise from tiny tongues of tissue growing upward from the thicker layer below, the dermis. These ridges form a rough surface that creates friction when it makes contact with other objects, in contrast to the skin over the remainder of the body, which is generally smooth. These so-called friction ridges have a profound evolutionary significance, reaching into our primate roots, because they enhance humans' ability to grasp objects.

Fingerprints, used by law enforcement since the end of the nineteenth century as a means of identification, are impressions left by residue from friction ridges on objects that have been touched. The most obvious fingerprints are those left on sticky or soft surfaces, such as dust, soil, blood, and paint. Visible to the unaided eye, these fingerprints often can be photographed for comparison with samples taken from suspects.

LATENT PRINTS

The type of fingerprints most frequently encountered during investigations, however, are latent, that is, not obvious to the eye unless artificially enhanced. Latent fingerprints are left not so much by a physical disturbance of a surface—like that of a thumb stuck into wet paint—as by secretions of glands in the ridges that produce sweat, amino acids, lactic acid, sodium, potassium, and glucose. Sweat evaporates quickly but the other secretions remain, and they create an impression that can be detected by forensic scientists.

Latent fingerprints are usually recovered in the forensic laboratory, although sometimes they are developed at the scene by dusting, a procedure best used when fingerprints are fresh and wet with sweat. Powder, colored

Above: Dusting for prints happens in real-life as well as the movies; the dust color is selected to highly contrast with the item being tested, for optimum visibility. Top left: In addition to a signature, a thumbprint further confirms a person's identity in this notary book.

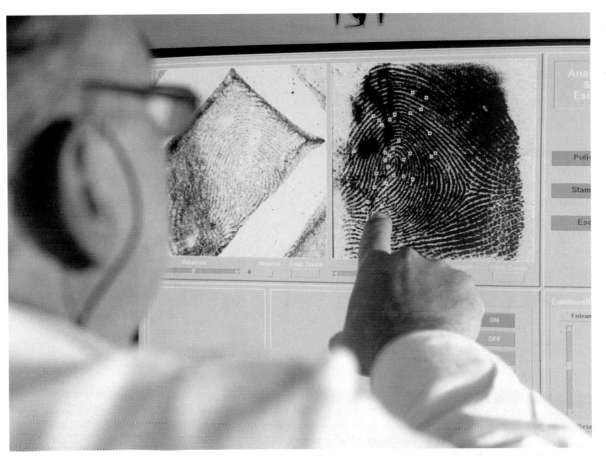

Forensic scientist pointing at fingerprints being compared by characteristic features, indicated by the square yellow markers. Fingerprints are classified into three basic types and are then further classified by features, which are called minutia points or ridge characteristics.

to contrast with the surface being tested, is brushed over the fingerprint. The powder grains adhere to the ridgelines, and the image is either photographed or lifted on adhesive tape, which is pressed on the fingerprint sticky side down.

Once sweat has evaporated from fingerprints, other processes are required to make them visible. One process employs superglues. Chemicals in these glues are attracted to the amino acids found on latent fingerprints. Typically, the glue is heated into a gaseous state within a chamber containing the materials being examined for prints. The gas adheres to the fingerprints and solidifies, outlining the prints in white. If glue-treated fingerprints remain obscure, they can be dipped in fluorescent dye, and then exposed to special lights, a process that brings them to life in vivid yellow, at which point they are easily photographed. A chemical called ninhydrin, which reacts to amino acids, also makes prints appear. Fingerprints sprayed with the compound appear in deep colors such as violet. Silver nitrate is another spray that works under ultraviolet light by bringing out salts left over from sweat. The most advanced fingerprint laboratories sometimes use lasers, which can make prints luminesce.

The fingerprints of suspects in criminal cases, corpses of unknown identity, and occasionally applicants for jobs requiring security and other special checks are recorded by inking the undersides of the fingers and taking the resultant impression. These fingerprints can then be matched against those from crime scenes kept on file by law-enforcement agencies.

Fingerprint Characteristics

During the nineteenth century, Czech physician and physiologist Jan Evangelista Purkinje, who later discovered sweat glands, studied fingerprints. Purkinje recognized that fingerprints could be classified according to the patterns formed by their ridges. These patterns, still used to categorize fingerprints today, are the arch, the loop, and the whorl. The prints of each person, so far as science can determine, can be judged by fingerprint experts around the globe using these three categories.

Of course, each of these classes can be subdivided further. An arch may be plain or tented: A plain arch is defined as a ridge that traverses the finger, from one side to the other, and has a hump in the middle, while a tented arch is a peak that is sharper than the normal arch, reaching an angle of 45 degrees in at least one ridge. A whorl has one or more ridges that form a circle, a full 360 degrees, centered at the middle of the fingerprint. A whorl could be oval, spiral, elliptical, circular, or double. A loop, as its name implies, has at least one ridge that starts on one side of the finger, loops back, and returns to the same side at which it started; depending on whether they are right loop or left loop, loops begin on one side of the finger (either right or left) and end on the other, and are the most common class of fingerprints, with arches the least common.

Left: Czech medical researcher Jan Evangelista Purkinje (1787–1869) made several pioneering studies, including histology, physiology, pharmacology, visual phenomena, and fingerprints. Top left: Significant details from a bloody fingerprint are noted.

MINUTIAE

Based on ridge patterns, fingerprint experts can match people to a distinct category, which narrows down a search for their identity but by no means creates a specific match. To achieve that, the experts rely on tiny characteristics, called minutiae, which essentially are variations and irregularities in the continuity of ridges. Minutiae, discovered at the end of the nineteenth century by British fingerprint pioneer Sir Francis Galton, include Y-shaped bifurcations of ridge lines, like splits in the road; sudden endings in lines; crisscrossing lines; lines that come to dead ends; and specks, like islands, between lines.

Experts differ on exactly how many minutiae can be categorized. Some count almost 20 different types. Others group several types of minutiae into just a few categories. However experts categorize fingerprints, evidence shows that the average print has about 150 different minutiae. The essence of considering an individual's fingerprints unique to that individual is that no two people have ever been scientifically proven to have exactly the same set of minutiae. As far back as 1892, moreover, Francis Galton calculated that at least 64 billion different fingerprints are possible, a figure that makes it highly unlikely that two living people have fingerprints that match in every detail. Not even identical twins have ever been found to have specific matches.

A match is made by comparing minutia points on inked fingerprints to the minutia points of fingerprints taken at the crime scene. Exactly how many of these comparison points must match to prove identity is something of an open question. The legal system in the United States generally recognizes the required number to be 12 points, as recommended in the FBI's fingerprinting manual. Fewer points are sometimes accepted, however, and in some cases it takes merely the testimony of an expert in fingerprinting to certify the match. European courts are stricter, mostly setting the number at 16 points, sometimes more. Yet, no matter how many points match, just one blatant difference in minutiae between two fingerprints, if it cannot be explained, can be enough to rule out the possibility that they came from the same person.

Law-enforcement agencies around the world have millions upon millions of fingerprints available on file for matching. For generations, these were kept in physical form, cards being the most common method. Today, records are automated, with vast computer banks of fingerprints accessed with electronic speed.

Francis Galton (1822–1911) developed the identification system that became the basis for the classification of Sir Edward R. Henry, who later became chief commissioner of the London metropolitan police. The Galton-Henry system soon spread across the globe.

When categorizing fingerprints according to their overall patterns, some individual's fingers do not fit into any of the classes, and some may have attributes of more than one class; an indicator can be innate or develop later in life, such as the scar left after a cut has healed.

Foolproof Fingerprints?

In April 2004, the Federal Bureau of Investigation took attorney Brandon Mayfield from Portland, Oregon, a Muslim convert, into custody in connection with the Madrid terrorist train bombings, which had killed 191 people a month earlier. The arrest of Mayfield, who had earlier represented a suspected Muslim terrorist in a child custody case, came after the FBI linked him to a fingerprint found at the site of one of the bombings. Two weeks later, the FBI revealed that Spanish authorities had matched the same fingerprint to an Algerian man.

The case is one of several that have been used to support arguments by critics of the fingerprint system who challenge its authenticity as absolute proof of identify and the claim that no two individuals anywhere in the world have fingerprints that are exactly alike. The challenge has gained the attention of both the news media and scientific publications. In September 2005, the respected magazine *New Scientist* asked, "How Far Should Fingerprints Be Trusted?" and in May 2004, the *Washington Post* published an article headlined "The Achilles' Heel of Fingerprints."

The argument has been waged within both the legal and scientific communities. The dispute hinges

Above: Fingerprint matches are not always foolproof. Oregon attorney Brandon Mayfield was taken into custody because a fingerprint in Madrid allegedly linked him to the train bombing. The FBI later apologized for the mistake. Top left: Fingerprints are taken with black ink on white paper for the greatest contrast and clarity.

on one issue that even the most ardent advocates of fingerprints as a surefire method of identification cannot deny. In point of fact, the uniqueness of fingerprints is a hypothesis that has never been proven empirically.

PROOF-POSITIVE PRINTS

Although matching fingerprints never have been found, no one can say for sure that no two people on Earth possess identical prints. The chances of finding two individuals with matching fingerprints, even if such a circumstance did exist, would be infinitesimal; nevertheless, because the contrary has not been proven, they may exist. Traditionally, courts have viewed the opinion of expert fingerprint examiners, combined with analysis showing that two sets of fingerprints match, as evidence enough to prove identity. Defense attorneys, with varying success, have challenged that hallowed concept.

Many such challenges are based on a 1993 case involving a pharmaceutical company, in which the Supreme Court ruled that courts must consider the relevance and validity of scientific evidence before it is admitted. This decision has fueled the argument about whether fingerprints are unique, but for the most part judges accept the presumption that acknowledged experts can testify to the 100 percent accuracy of fingerprints that they have examined and identified. There is, however, a caveat. Even law-enforcement agencies admit that mistakes can be made, as apparently happened in the Mayfield case. The question still unanswered is whether a mistake by the examiner while analyzing fingerprints makes the method of examination itself open to challenge. In the Mayfield investigation, the FBI found 15 similar comparison points, although the quality of the image of the fingerprint used for the comparison was later faulted. Spanish authorities, on the other hand, pinpointed only 8 for their suspect. One of the most vocal critics of the infallibility of fingerprints is Simon Cole, a criminologist at the University of California, Irvine. Cole claims that the percentage of false fingerprint matches is more than one in a hundred.

Whatever the outcome of the argument, it is likely that this will lead to more research intended to verify, to scientific satisfaction, the uniqueness of fingerprints.

Just as a medical doctor seeks a second opinion for a diagnosis, a "positive identification" of a fingerprint is not accepted until verified by another certified expert in the field.

ART OR SCIENCE?

The outcome of fingerprint analysis depends to a very large degree on the expertise, experience, and opinion of the examiner. Thus, while it uses scientific techniques and is based on scientific theory, fingerprint analysis can be considered an art rather than a science.

The question has arisen whether an examiner can be prejudiced toward a certain outcome when making a match. In many cases, investigators are overtly seeking to match fingerprints from the scene to those of a suspect, even if unconsciously. The examiner usually knows this. In the instance of Brandon Mayfield, examiners used by the FBI knew that Mayfield had converted to Islam and had represented a client with alleged terrorist sympathies. Analysis requires that an examiner draw a conclusion as to whether or not fingerprints match, even if the specimen is damaged or only a partial print. Issues such as these are very much in the minds of defense attorneys when fingerprints are presented as evidence against their clients.

Ballistics

Above: To determine whether a gun was involved in a specific shooting, investigators test-fire the weapon and compare the small markings left on the bullets with those from specimens found at the crime scene. Top left: Due to the high decibel level of gunfire at an indoor range, ear protection should always be worn to preserve the tester's hearing.

Ballistics, like fingerprinting, is a staple of forensic science, and unlike areas such as genetic profiling, it is one that relies on concepts, techniques, and equipment little changed since first employed in crime laboratories in the 1920s. "Ballistics" is a term that refers to projectiles and firearms. It can be defined as the study of projectile dynamics, the flight characteristics of projectiles, firearm functioning, and the flight and effects of ammunition. When firearms are recovered from a crime scene, ballistics plays a major role in successfully concluding investigations by matching bullets to the guns that fired them. Ballistic evidence is based on inherent qualities of firearms, cartridges, and bullets, the dynamics of bullets in flight, and the explosive reaction that occurs when a cartridge is discharged. Cartridges with self-contained components have been in use since the mid-nineteenth century, replacing firearms, such as those with percussion caps, in which the ignition, propulsion, and projectile were loaded separately. Analysis of cartridge components is a key part of forensic ballistics.

CARTRIDGE COMPONENTS

The anatomy of a cartridge is fairly simple. The projectile is the bullet held in place at the fore of the cartridge case, and the cartridge case is the component that holds powder, which, when it burns, produces gases at pressures high enough to propel the bullet down and out of the barrel. Contrary to what many people think, powder burns not explosively but progressively, creating a slow burn that propels the bullet at the particular velocity for which it is intended. Most rifle and pistol cases are brass or steel, while shotgun cases are plastic. At the rear of the case—actually called its "head"—is the primer, which contains explosive compounds that detonate on impact and ignite the powder. Most primer materials are in a cup at the center of the cartridge. Many low-caliber, low-power cartridges, however, have primer in the rim of the casing.

TYPES OF BULLETS

There are many different bullet styles. The basic bullet is a molded oblong lump of relatively soft lead, which may be hardened by alloying it with metals such as tin. Soft bullets deform on impact, spreading and fragmenting, causing wide areas of damage. The standard bullet is adapted for different uses by design and chemistry. Some bullets used in high-velocity firearms are jacketed in metals such as brass or synthetics such as nylon to protect the bore as they pass through it. These bullets are partly jacketed, leaving the nose exposed for deformation. A hollow point on jacketed bullets achieves a similar purpose. Fully jacketed bullets have a high degree of penetration and result in a significant mortality rate.

PISTOL OR HANDGUN?

People unfamiliar with guns often call all handguns "pistols." The firearms purist rejects this definition. Technically, a handgun in which bullets are contained in a rotating cylinder and are fired sequentially is a revolver. Other handguns are pistols, typically semiautomatic, although they can be single shot. A handgun in which bullets are delivered to the chamber from a spring-loaded clip, usually in the handle, at the pull of the trigger is a semiautomatic pistol. Empty casings are ejected during the process. Single-shot handguns also are known as pistols.

Rifles are long guns in which bullets can be loaded and unloaded, semiautomatically or manually, by cranking a lever or pushing and pulling a sliding bolt. Some rifles are single shot; a new bullet must be loaded after each shot is fired. Shotguns have similar "actions," as loading mechanisms are called.

Traditionally, many gun owners have loosely referred to semiautomatic firearms as "automatics." They have come to regret the nomenclature, however, because gun-control advocates and, often, the media, blur the distinction as well. An "automatic," properly, is a weapon that continuously fires as long as the trigger is depressed, or until ammunition is depleted. Possession of such firearms is tightly regulated by federal law and banned by some state and local authorities. An easy way to picture a fully automatic firearm is to think "machine gun."

A hunter loads his rifle. Each type of firearm holds cartridges of a specific caliber. This fact, and an individual firearm's trademark rifling, link bullets to the gun that fired them.

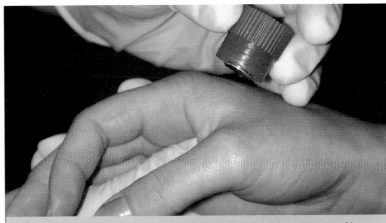

A forensic expert can test for gunshot residue of lead, barium, antimony, or a combination of the three on the hand of a suspected shooter, which indicates that a gun was fired.

GUNSHOT RESIDUE

The detonation of a primer and powder in a cartridge often leaves a residue of chemicals and particles on the victim, the assailant, and the crime scene itself. This material, called gunshot residue, or, more succinctly, GSR, can provide clues detectable by forensic scientists. GSR must be collected quickly because it is easily removed by water, wind, and wear.

Several methods are used to examine GSR, including treatment with chemicals that react with heavy metals in the residue, and scanning by neutron activation analysis or electron microscope, which provides magnification that reveals details on the surface of microscopic particles, to compare with known samples. GSR testing has come under attack by some critics, however, who complain that the process is not standardized.

Firearms Comparisons

The bore, the inside surface of a handgun or rifle barrel, is marked with rifling, spiral impressions from grooves cut into the metal. The raised portion of the rifling, above the grooves, is called land. Rifling creates friction that spins a bullet as it passes through the barrel, stabilizing it, which is necessary for an accurate flight path, which would otherwise be wobbly due to the bullet's oblong shape. The long guns of yesteryear, such as muskets, were made with smooth bores, the reason they were much less accurate at long ranges than rifles. Shotguns also have smooth bores, but because their shells contain pellets that spread into wide patterns, they do not depend on pinpoint accuracy to hit their mark.

Rifling leaves markings called striations on a bullet in the form of lines parallel to its long axis that vary from one type of firearm to another, in width, direction of twist, and number of lands and grooves. A pistol made by one company may have six grooves that twist to the left, while one of the same caliber from another manufacturer may have rifling with a twist to the right. Forensic ballistics technicians examine these characteristics and compare them with guns from known manufacturers to ascertain the make of gun involved in a crime.

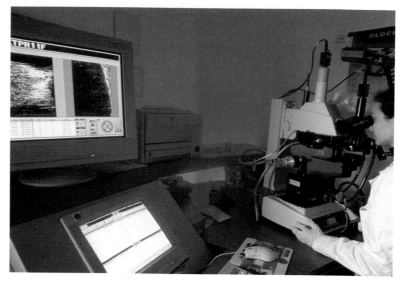

Above: Each gun's rifling leaves its own distinctive pattern on the bullets it fires. These markings can therefore be examined under a comparison microscope and a reliable match made. Top left: Automatic weapons eject their shells after firing, often leaving casings littered about a crime scene.

BARREL "PRINTS"

Each barrel, moreover, is unique in that it contains imperfections in the metal left during the manufacturing process, a condition that gunmakers obviously try to keep to a minimum. Rifling machinery may become worn with use or work with less-than-desired uniformity from one gun to another. A gun's rifling itself can also wear down after repeated firing. The imperfections leave marks on every bullet fired from that weapon, making it possible for examiners to link bullets to specific firearms. As with fingerprinting, matching bullets to guns is a matter of painstaking comparison. The firearm is test-fired into a target such as water, which cushions the bullet. Next, the striations on the fired bullet are compared with those on the bullet from the crime scene. This process takes place under a comparison microscope, essentially a pair of microscopes bridged so that two objects can be viewed simultaneously at the same degree of magnification. If both bullets were fired from the same gun, they will match.

Guns also uniquely mark cartridge casings so that if one or more of these are recovered at a crime scene, they can link a suspect's weapon to the crime. As are bullets, casings are examined for markings under a comparison microscope. When the trigger is pulled and activates the firing pin, it strikes and dents the bullet's primer. The impression left by the firing pin is identifiable. Other markings occur when the powder in the cartridge explodes on discharge, with the firing chamber making marks on the back of the cartridge; these markings also can be matched to the firearm. Automatic and semiautomatic firearms have mechanisms called the extractor, which pulls the empty cartridge from the chamber and the ejector, which expels the cartridge from the gun. Both of these mechanisms produce their own distinctive markings, usually at or near the rim, which can be identified by examiners.

An automatic is so named because the recoil of the gun sends a toggle backward and upward to eject the spent cartridge and chamber a new one from a magazine.

Just as a gun leaves distinctive marks on the bullets it fires, it also makes its mark on the expelled casing of the bullet, so that investigators can match the casings with a specific gun.

Trace Evidence

The tried-and-true principle that a criminal will always take something from a crime scene and leave something behind means that no tiny trace of evidence— not a barely visible chip of paint, a fiber too delicate to be retrieved by hand, or a hairline crack—can be disregarded when investigators sweep a crime scene. All this trace evidence has the potential to link people, places, and things, and ultimately resolve the question of criminal responsibility.

TRACKING PAINT
Unless a criminal does something as foolish as stepping into wet paint—it does happen—paint evidence is usually in the form of bits and chips of dry paint. The crimes that most frequently produce paint evidence are those involving vehicles, such as hit-and-runs or getaways that involve even minor collisions. Examiners analyze paint samples microscopically and with techniques such as gas chromatography to determine their chemical constitution and, if possible, their manufacturers. They may look for characteristics that indicate the paints are specialty items, with a limited number of manufacturers and buyers, narrowing down the search. Law-enforcement agencies keep catalogs and data banks of identified paint samples for comparison with evidence. The kind of paint can indicate the make and model of a car, and even its year of manufacture.

MATCHING FIBERS
Fibers can come from a huge number of items in everyday use—clothing, towels, floor mats, seat covers, and paper, to name a few. Fibers can come from natural sources, such as wool or cotton, or can be created by humans. Human-made fiber sources include synthetics, such as nylon, and regenerated fibers manufactured from cotton or wood pulp. Unlike fingerprints

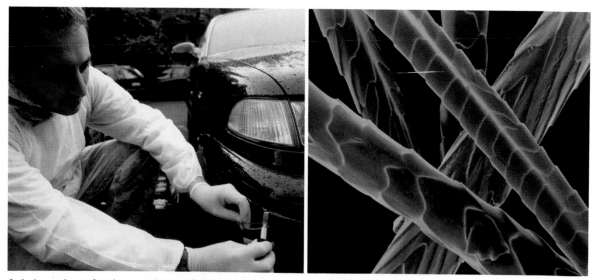

Left: A scraping tool retrieves a paint sample from a car hit by a get-away vehicle to garner information about the other auto's possible make and model. Right: When looking under a microscope at fibers such as the angora wool pictured, scientists can often match these strands with materials at the crime scene. Top left: Crime-scene investigators comb a crime scene for trace evidence.

Glass can prove powerful evidence in a case, since the type of glass, as well as the way in which it was broken, can help trace and re-create the perpetrator's path.

chemical composition. Matching usually is attempted with materials at the crime scene, not from a fiber collection, because no extensive records of fibers exist.

GLASS ANALYSIS

Whether from a windowpane broken by a burglar, a bottle smashed over a skull in a bar fight, or a shattered automobile windshield, glass makes up important evidence in many criminal investigations. Glass has physical, optical, and chemical properties that can differ according to the way in which it was made and the purpose for which it is intended. By identifying these properties, examiners can trace glass to its origins. The kind of glass used in an automobile headlight, for instance, can be tracked to the car's make and model.

Glass fractures in different ways according to the kind of impact that broke it. A gloved fist smashed into a window, for example, causes a different kind of break than a bullet does. Analysis can help re-create the circumstances surrounding a crime.

Imagine that a burglar upsets a glass lamp, which falls and shatters on the floor. As he moves about the room, a tiny shard from the lamp sticks to his shoe. After the burglar is apprehended, examination of his shoe reveals the shard. Examiners can show that the shard came from the lamp by matching its fracture lines with those of pieces found in the room, much like putting together the parts of a jigsaw puzzle. It is a painstaking process, but it works.

or rifling marks on a bullet, fibers are not conclusive evidence, only indicators of conditions such as their presence on the person of a suspect and their source in a piece of furniture at a crime scene. Examiners try to determine the size, shape, and conformation of fibers under a microscope, as well as their

Questioned Documents

In the eyes of law enforcement, the definition of a questioned document goes a long way beyond ransom notes and forged checks. Included among documents that are examined for clues to crimes is virtually anything upon which a mark has been made for transmitting a message, including graffiti, inked impressions from rubber stamps, and mechanical and automated check writers.

Document examiners do much more than analyze handwriting and typewritten script. Their duties include historical dating; analyzing paper, ink, and watermarks; determining whether documents are forged or doctored; examining computer printouts; and reconstructing burned and otherwise damaged documents.

MAKING THE CONNECTION

Documents can link a suspect to a crime in several ways. A note handed to a bank teller by a robber can be matched with handwriting found in the suspect's home. Type can be matched to that of a typewriter in the possession of a suspect. A file recovery engineer can determine that a computer file in a child pornography case has been deleted from a suspect's personal computer. It takes a trained eye to discover flaws in documents;

Above: Looking under a microscope, an investigator can use bright white or ultraviolet lights to determine if symbols such as watermarks are false. Often tiny details too difficult to fake are visible only through this lens. Top left: Forged checks can easily link a suspect to a crime.

just the manner of dotting an *i* or crossing a *t* can determine the difference between a legitimate check and one that is flawed. In one case investigated by the Georgia Bureau of Investigation (GBI), a suspected check forger was asked to write the victim's signature three times.

Although the suspect carefully attempted to disguise her handwriting, the examiner used several characteristics, including the connections between the letters *u* and *d* in the surname "Catudal," to link the sample to the signature on the check.

ELECTRONIC DEVICES

Document examiners use not only their experienced eyes to find flaws but instruments as well. A device called the infrared electronic converter can detect differences in inks that appear identical to the naked eye. With the use of this handy tool, inks can glow, vanish, or appear as they

were before illumination, depending on their chemical properties. One type of instrument used by document examiners is a projector that compares machine impressions from devices such as typewriters, check writers, and notary seals. It can determine whether two documents were made by the same device through superimposing one reflected in red light on the other reflected in green light. Since red and green light combine into black, the color of letters precisely aligned, one over the other, appear black.

FIGURING OUT A FORGERY

Handwriting analysis is a true art, especially since interpretation is often in the eye of the beholder. While examining a

forgery, an analyst may look for qualities in the writing that suggest that the writer is trying too hard: script that is laborious, shaky, and irregular; marks that indicate that the writing instrument has been frequently lifted off the paper, perhaps so that the writer can rest his or her hand; heavy-handed crossing of *t*'s and dotting of *i*'s and heavier instrument pressure in general; a signature that appears too formal; and inconsistent spacing after capital letters.

Handwriting analysis, as noted, is subject to interpretation and requires a keen eye and extensive knowledge and experience. In the end, it is up to the court to decide on the weight of the examiner's expert opinion.

Left: Using a keen eye, investigators can discover signature discrepancies. Crossed t *'s, dotted* i *'s and irregular, shaky script are only some of the many ways they interpret forgery. Right: From old-fashioned typewriters to new check writers, each machine leaves its own distinctive record.*

Advances in Forensic Science

600s BCE

Chinese possibly use fingerprints for personal identification.

35–95 BCE

Quintilian uses bloody palm print as evidence in court.

1248

Xi Yuan Ji Lu, which is loosely translated as "The Washing Away of Wrongs," is published in China by Song Ci.

1561

Ambroise Paré publishes book on human anatomy.

Ambroise Paré

1600s

Paolo Zacchia and Fortunatus Fidelis lay the foundations of modern pathology.

1686

Marcello Malpighi, a professor at the University of Bologna, first uses a microscope to find fingerprint patterns.

1775

Swedish chemist Karl Wilhelm Scheele discovers a method of detecting arsenic in body.

1806

Valentin Ross extracts arsenic from body in a detectable form.

1813

Mathieu Orfila, Spanish-born professor at the University of Paris, publishes a major book on the nature of poisons and how to detect them; he founds modern toxicology.

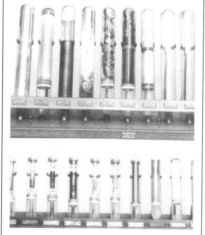

Arsenic and other distilled chemicals

1823

Czech Jan Evangelista Purkinje, anatomy professor, publishes book on fingerprint patterns, although he does not link them to individual identification.

1832

James Marsh develops a method that detects arsenic in other materials, even in small amounts.

Mathieu Orfila

1856

William Herschel, a British official in India, first uses thumbprints of people in his jurisdiction to sign contracts.

Jan Evangelista Purkinje

1880

Henry Faulds, a Scot working in Japan, publishes an article describing how fingerprints could be used as a form of personal identification.

1882

French policeman Alphonse Bertillon introduces anthropometry, a systematized use of body measurements as identification.

1893

Austrian professor Hans Gross publishes *Criminal Investigation,* the first comprehensive treatise on the application of scientific disciplines to crime solving.

1889

Alexandre Lacassagne, French forensic scientist, recognizes lands and grooves, the distinctive marks made when a bullet passes through a rifle barrel.

1892

Francis Galton publishes the ground-breaking book, *Finger Prints,* which establishes fingerprints as a means of personal identification and notes their individuality and permanence.

1896

Edward Richard Henry, a British civil servant in India, establishes groupings of fingerprints and a classification system for law-enforcement use.

Various fingerprint patterns

1901

Edward Richard Henry becomes head of the first Scotland Yard fingerprint bureau.

1901

Karl Landsteiner discovers blood groups: A, B, O, AB.

Blood

1910

Edmond Locard develops concept of transference of trace evidence, leading to formulation of Locard's Exchange Principle.

1910

Edmond Locard opens the world's first crime laboratory in Lyons, France.

1910

In his book *Questioned Documents,* Albert S. Osborn sets standards for handwriting analysis.

1915

Italian Leone Lattes converts dried bloodstains into a liquid form suitable for testing.

1925

U.S. Army colonel and physician Calvin Goddard publishes an article describing the use of the comparison microscope to match spent bullets by their lands and grooves.

1916

August Vollmer opens the first crime laboratory in the United States, in Los Angeles.

1932

The United States Federal Bureau of Investigation opens its first crime laboratory.

J. Edgar Hoover

1933

The United States Bureau of investigation is renamed the Division of Investigation.

1977

The Federal Bureau of Investigation introduces Automated Fingerprint Identification System.

1985

British geneticist Alec J. Jeffreys announces the development of first DNA profiling test, which has had profound impact on legal systems worldwide.

DNA strand

Becoming a Forensic Scientist

Forensic scientists work in the world of law enforcement, but they are first and foremost scientists. Therefore, no matter what specialty field a would-be forensic scientist aspires to, he or she needs academic credentials in the particular science or sciences involved.

The Forensic Sciences Foundation, an affiliate of the American Academy of Forensic Sciences, provides guidance to individuals wishing to join the ranks of those who use science to fight crime. Their general advice, adapted here, indicates that a potential forensic scientist should have the following credentials and skills:

- A bachelor's degree in science, and often an advanced degree, or degrees, in a specialization,
- Good speaking skills,
- Good note-taking skills usable in real-life situations,
- The ability to write an understandable scientific report,
- Intellectual curiosity,
- Personal integrity.

Forensic computer analysts, for example, should have, at minimum, certification or a bachelor's degree in information technology, and there are now master's programs in the field. Forensic odontology demands a doctor of dental science degree as a basic requirement. To gain necessary expertise, however, candidates need to complete specialized courses in areas such as medicolegal death investigation. Forensic pathology is a specialty for medical doctors; thus, a doctor of medicine or doctor of osteopathic medicine degree is a mandate, as is additional study in forensics and the standard internship and residency required of all physicians. People who hold Ph.D.s in certain biological fields also can become forensic pathologists.

The Criminalist

Accounts of police activities reported in the news sometimes refer to "criminalists" investigating a crime. The term is sometimes misunderstood. Some people think of criminalists as investigators, even detectives, or else as scientists who are involved in researching the activities of criminals. In

Above: Because they might work with potentially dangerous substances and chemicals, criminologists in the laboratory wear protective clothing and devices such as gloves, masks, and goggles or eye protectors. Top left: As with most sciences, forensics requires a college education and even advanced degrees.

reality, a criminalist can be a bit of both. Essentially, a criminalist is a forensic science technician and is often described as such in the labor market. The job of the criminalist is to identify, compare, and interpret physical evidence that can link a particular suspect to a specific crime and to the victim. A criminalist can work both at the crime scene and in a laboratory.

One of the major job functions performed by a criminalist is to distinguish between materials that are potentially valuable as evidence and those that are not of consequence. Another skill required of a criminalist is the ability to describe the results of evidence analysis in written form and verbally in a manner understandable in court. Criminalists often must take the witness stand.

Criminalists are usually bench scientists, generalists who perform basic laboratory testing, such as serological or impressions analysis, which may or may not then be passed up the chain of command to scientists with greater expertise in individual fields. Generally, an entry-level criminalist must have a bachelor's degree in a science such as chemistry, physics, or biology or in criminalistics itself, a major offered at several colleges and universities. Many criminalists have graduate degrees as well.

Forensic Accounting

Accountants do not have a particularly exciting public image, but they can play a major role in bringing criminals to justice. The FBI and other major law enforcement agencies have long hired agents with a background in accounting because of their analytical skills; records of financial wrongdoing have often been employed to put criminals behind bars. After all, Al Capone's tax problems,

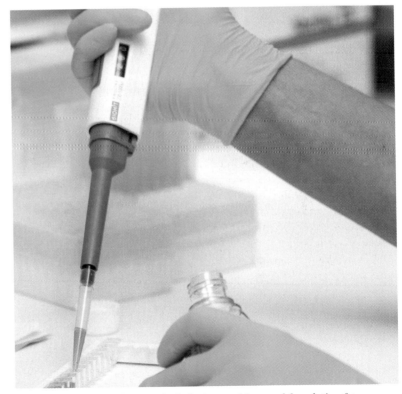

Above: A chemistry, biology, or physics background is a good foundation for pursuing a career in criminology. Top right: Although the investigative and analytic aspects of criminology may take the forensic scientist to a crime scene or the lab, criminologists are also often called upon to testify in court.

rather than murders and boot-legging, sent him to prison. Today, forensic accounting has gained even more importance as law enforcement seeks to prosecute white-collar crime such as fraudulent financial reporting. Forensic accountants typically have been used to expose money laundering, an aspect of law enforcement even more important when terrorists use it to finance their activities. Although forensic accounting is not new, the stress on "forensics" in schools that train people for this purpose is a growing trend. Along with accounting studies, would-be forensic accountants take courses that teach techniques they will need as investigators and expert witnesses.

Forensic Animation

Computer-generated animation of an alleged crime or events surrounding one is a new method of presenting evidence in court, and although it shows great promise, it is not without controversy. Forensic animation came into favor during the 1990s, first in civil cases, and then in criminal investigations. Essentially, the technique is a computerized animated illustration of evidence, both physical and by means of the interpretation of testimony. In a murder case, for example, this visual aid can demonstrate the trajectory of a bullet, as determined by investigators,

passing through a victim's body. Prosecutors have found that computerized animation can have a great impact on juries. A problem raised by some defense counsels, however, is that animation is only a re-creation of hypothesized evidence: it is not a recording of the real event. From a defense standpoint, the concern exists that jurors may confuse the two, as if mistaking a video game for real life.

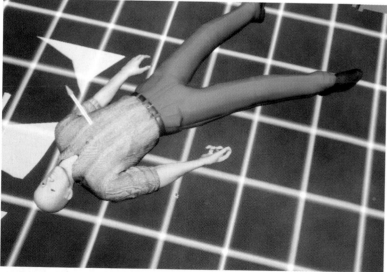

Above: Forensic animation may assist a judge or jury to visualize a crime. Defense attorneys fear the suggestive nature of these "re-creations," because people are more apt to believe something they see—even though it might only be a reenactment of theorized evidence or even conjecture. Top right: A forensic nurse may be called on to treat victims of a crime, assist an attorney with legal issues, and testify in court.

Forensic Nursing

In a field that already offers a multitude of subspecialties, forensics can now be added to the list of areas nurses can choose to concentrate in.

Forensic nurses perform the many different types of nursing work associated with legal issues and disputes. They counsel lawyers on nursing and medical issues in cases, supply expert testimony in court, and provide nursing services to victims of crimes, often victims of sexual or domestic assault. These nurses work in hospital emergency rooms with crime victims, obtaining patient histories and descriptions of the reported incident, performing physical exams, documenting the physical and emotional condition of the victim, and collecting physical evidence. If a case comes to trial, the forensic nurse who examined the victim often gives expert testimony about the exam results. Some forensic nurses help set up sexual assault response teams in hospitals or police departments or do educational programs on sexual assault. Still others work with criminals in prison. Besides being either a licensed or registered nurse, most forensic nurses need additional training provided by their employer or a professional organization such as the International Association of Forensic Nursing.

FIELD SAFETY

Crime has its hazards. So does investigating crime—and not only the many risks involved in dealing with dangerous criminals. Forensic investigators often risk exposure to potentially dangerous situations and hazardous materials. Pathogens, toxic chemicals in a variety of forms, radiation, airborne contaminants, confined spaces such as caves and sewers—these and myriad other risks may confront investigators.

Because investigators commonly come into contact with blood and other body fluids, as well as weapons and hypodermic needles contaminated with them, they need protective equipment such as disposable gloves and coveralls, and sometimes goggles and face masks. "Sharps"—a medical and forensics term for items such as needles and knives—must be handled with great care and stored in containers that are puncture-proof. Investigators must mind their personal actions as well. Even when one is wearing protective clothing, it is ill-advised to eat or drink where infectious materials are present. Nor is it wise to apply cosmetics in such an environment, mandates the FBI's *Handbook of Forensic Services*.

The variety of special gear used to protect crime-scene investigators is extensive: respirators when airborne contaminants are present; body harnesses and backup communications when entering sewers, pits, wells, and other dangerous spaces; and full-body hazmat (hazardous material) suits in cases where biohazards or radioactivity might be present. The mounting danger of weapons of mass destruction has made it necessary for major law-enforcement agencies to establish specially trained teams to investigate events in which these may be involved.

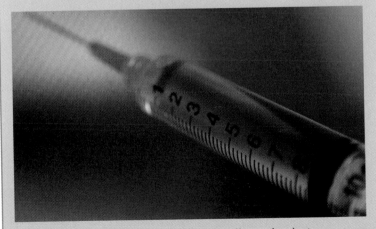

In an effort to fight infection by contaminated needles or other sharp instruments, investigators in the field utilize protective garments and gear.

Forensic Science Specialties

The study of many scientific disciplines and other subjects can lead to the field of forensics, and the number of disciplines only increases with the advent of further developments in the science. Here are some of the disciplines, including some that are not specifically scientific in nature, that may play a major role in criminal investigation:

- Accounting
- Anthropology
- Art/Animation
- Ballistics
  (involves physics)
- Biology
- Botany
- Chemistry
- Computer analysis
- Dentistry (Odontology)
- Document Examination
- Engineering
- Entomology
- Geology
- Meteorology
- Microbiology
- Nursing
- Palynology
  (study of fossil pollens)
- Pathology
- Photography
- Psychiatry
- Toxicology

Areas of Study

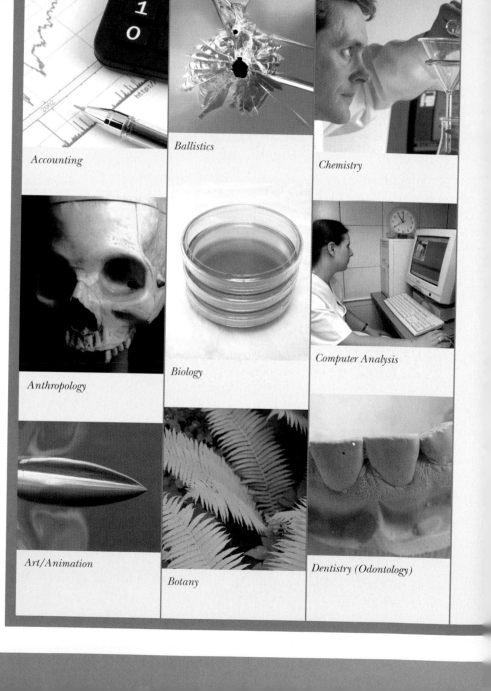

Accounting

Ballistics

Chemistry

Anthropology

Biology

Computer Analysis

Art/Animation

Botany

Dentistry (Odontology)

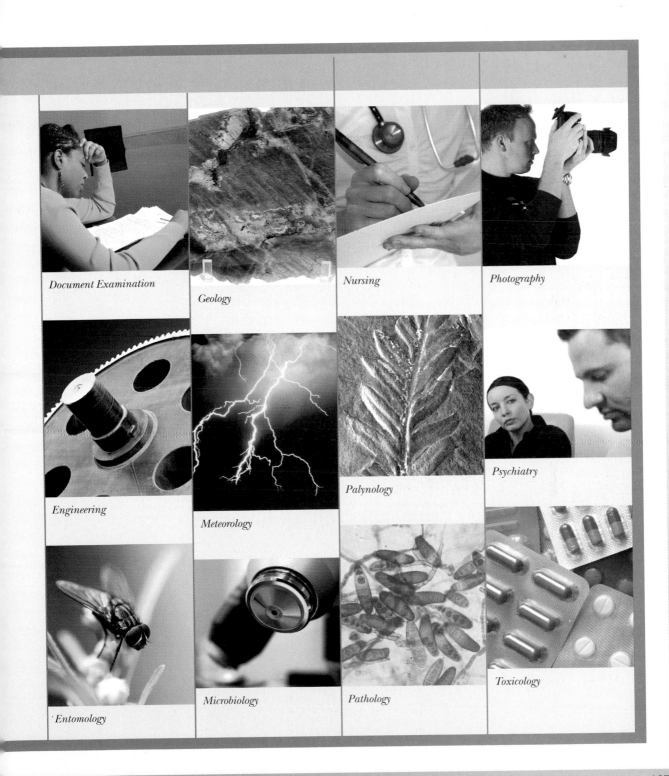

Document Examination

Geology

Nursing

Photography

Engineering

Meteorology

Palynology

Psychiatry

Entomology

Microbiology

Pathology

Toxicology

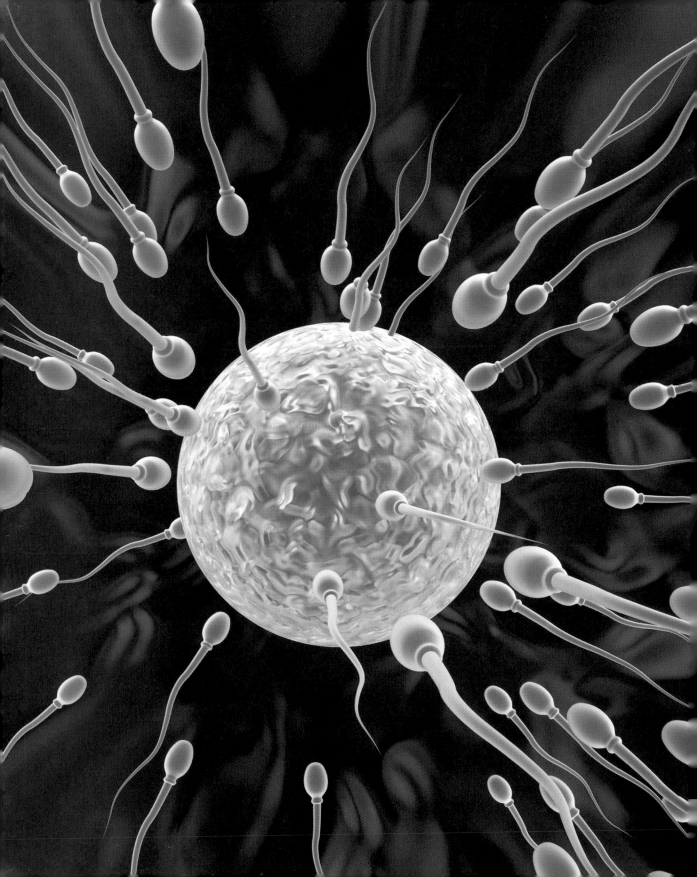

CHAPTER 7

DNA AND MICROBIAL FORENSICS

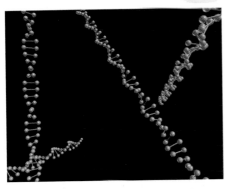

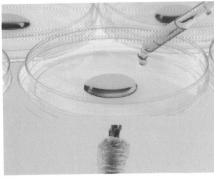

Left: As soon as the sperm and egg cell unite, all the inherited characteristics of a person's makeup, such as eye shape, hair color, or size of the ears, are determined. Top: DNA is an organic compound present in every cell of a human body and includes necessary genetic information. Bottom: While scientists have decoded a large amount of the DNA molecule, there is still much research ahead of them.

From the moment sperm and egg cells unite, parents pass along the greatest inheritance they can gift to their offspring. It is the gift of self, of what makes one person's eyes blue and the other's brown, of what gives an individual that certain smile, or makes one tall and another short. That gift is incorporated in an organic compound called DNA, which resides in every one of the body's living cells. As organic chemicals go, the DNA molecule is not particularly intricate, but then on anyone's scale all molecules may be considered complex. Scientists have worked for years to unravel the DNA molecule and have it fairly well figured out, although by no means completely. Study of DNA has revealed vast amounts of information about human genetics. It also has made this carrier of genetic traits a remarkable new tool in the hands of forensic scientists. The use of DNA profiling today reflects the way in which the scope of forensics has grown out of new developments in science generally, a trend that began long before the forensic sciences were recognized as an organized body of knowledge with crime-solving applications. The use of forensic DNA technology, in turn, fuels public enthusiasm for further DNA research.

About DNA

The acronym "DNA" has become such a byword that the term it stands for is often forgotten: deoxyribonucleic acid. Whether you use the acronym or its full name, this chemical compound is the guiding force that shapes who and what you are. You have your mother's eyes and your father's ears—and perhaps a hot temper from one of them—because of DNA inherited from them. What parts of the parental DNA you receive can be said to be a potluck affair, because DNA is passed on in relatively random fashion.

WHAT IS DNA?

DNA is sometimes called the "blueprint for life." DNA issues orders for tasks to be performed to living cells in the body, which amount to a considerable number of cells. The estimates of how many cells the average human body contains vary greatly, from approximately 10 trillion to as many as 100 trillion. Whatever the exact number may be, within that range, DNA has a monumental number of workers under its supervision. DNA tells cells not only what to do but also when to do it.

DNA is found in the chromosomes, the long strands that contain genes, within the nucleus of living cells. The term

Above: DNA inherited from parents carries the genetic information that produces what is known as "family traits," including the color of eyes, the shape of an ear, and even our smiles. Top left: A strand of DNA, which is often described as resembling a ladder, rope, or chain.

"double helix," describing what the DNA molecule looks like, is not the easiest to visualize without a picture for reference. Less technical ways to describe it are as a twisted ladder, rope, or chain, or a spiral staircase. The molecule belongs to a group of chemical compounds called polymers, which resemble chains and consist of links, sometimes millions of them, repeated again and again. The rungs in the ladder or, if you will, the stairs, are known as nucleotides, or

bases, and in DNA there are four different types, often referred to for simplicity by the letters A, G, C, and T, for adenine, guanine, cytosine, and thiamine. Each rung is a pair of bases joined together, totaling about 3 billion pairs in the human body. A tiny length of DNA can contain millions of bases, joined together in virtually any number of combinations and sequences that can be imagined. The way these bases are combined in DNA writes the code for what we are, just as letters of the alphabet can be connected to form a myriad variety of words.

Each group of living organisms shares a considerable amount of DNA; more DNA is shared as taxonomical relationships become closer. As primates, humans and lemurs have a large amount of DNA in common; chimpanzees and humans share about 95 percent; and all humans share about 99 percent. In people, the remaining small percentile accounts for the differences between siblings, parents and offspring, neighbors and neighbors, and, in short, all of us as individuals. Only one-tenth of 1 percent of DNA differs from one person to another, and that is the portion forensic scientists use to develop DNA fingerprints.

Contrary to popular belief, it was not Watson and Crick, but Johann Friedrich Miescher, a Swiss specialist in cell biology, who, in 1869, discovered the chemical DNA.

WHO DISCOVERED DNA?

Who discovered DNA? If you answered James Watson and Francis Crick, you are wrong. Despite what you may have read or heard, they neither discovered DNA nor made the connection that it was part of the human body's genetic machinery. They never claimed credit for doing so. What Watson and Crick did in 1953 was to discover the three-dimensional molecular structure of DNA and present it as a model, the famed double helix.

The chemical DNA itself was discovered in 1869 by Johann Friedrich Miescher (1844–895), a Swiss specialist in cell biology. By the middle of the twentieth century, scientists had learned that DNA was a genetic substance, and many researchers devoted their work to its study. Then came the breakthrough by Cambridge University researchers Watson and Crick. In the years immediately after their find was announced, writers, educators, and science journalists were generally very careful to define the exact nature of the Watson-Crick discovery. As time has passed, however, the specificity of just what Watson and Crick did has often been treated somewhat sloppily and referred to as the "discovery" of DNA.

Chimpanzees and humans share about 95 percent of the same DNA, and humans share about 99 percent with each other.

Discovery of DNA Fingerprints

DNA fingerprinting, also called DNA testing, typing, and profiling, has become the "glamour" area of forensics, so much so that in the mind of the public it often overshadows other tried-and-true techniques that have proven their worth over generations. Whatever the public perception of this procedure, and even if it is not the ultimate forensics tool, it assuredly is one of the most important developments thus far in the forensic sciences. DNA, or genetic, fingerprints play essentially the same role as actual fingerprints, which police have used since early in the twentieth century to identify suspects and victims of crime. DNA fingerprinting uses identifying characteristics of individual DNA samples to determine whether they match, thus indicating that they came from the same individual.

A RANDOM DISCOVERY PROVES PROMISING

DNA fingerprinting is a relatively recent development in forensic science. It was discovered in 1984 by British geneticist Alec J. Jeffreys of the University of Leicester. Jeffreys did not sit down one day and decide to find a way to identify people genetically. His discovery came after a chance development during research along an entirely different line. This circumstance shows how a brilliant scientific mind can see potential in an event that others might ignore.

Jeffreys was studying a particular gene, the one for myoglobin, a protein that enables muscle cells to store oxygen. Seals, whales, and dolphins have enormous amounts of myoglobin in their muscles, enabling them to remain underwater for extended periods of time. In fact, Jeffreys was examining seal myoglobin in hopes of pinpointing the same gene in people. He found that the gene contained short segments of DNA whose combinations and sequence of bases were unique to each individual. Working with colleagues, he studied these segments further and developed ways to purify and label them. The comparison of segments such as these between individuals is the basis for DNA fingerprinting.

DNA SOLVES ITS FIRST CASE

Jeffreys was soon able to put his findings to a practical test when chance circumstances thrust the procedure into the limelight. In the space of three years, two teenage girls had been raped and murdered around the small community of Narborough, not far from where Jeffreys worked. Police found type A blood at the scene of both killings, which also yielded semen samples. A local teen was questioned intensely and eventually confessed to one murder, but not the other. The boy's family questioned the confession, and police went to Jeffreys, asking that he perform his new DNA test on the two semen samples. The samples matched each other but not that of the suspect. It was obvious

Above: While studying the gene for myoglobin, which is what allows whales to stay underwater for extended periods of time, British geneticist Alec Jeffreys stumbled upon the basics of DNA fingerprinting. Top left: Samples for DNA testing can be taken from blood splattered at a crime scene.

A criminal may leave bodily evidence behind, such as blood, semen, or saliva, that are DNA "fingerprints." Tests can match sample to suspect.

that the same man committed both murders, and police had the wrong person. The boy was released. Police then initiated the world's first mass DNA sampling operation. They took blood samples from 5,000 men in the area. Once men with type A blood were isolated, they were subjected to DNA testing. None matched those of the killer.

Police were stumped by the findings. Then a local woman reported overhearing a colleague talking about how he had given a blood sample while claiming to be a friend with a surname that seems as if it were contrived for a murder mystery: Colin Pitchfork, a baker in the area. A sample of Pitchfork's DNA was then taken and found to match the semen from the crime scene, leading to his arrest, conviction, and a sentence of life in prison.

The case foreshadowed the use of DNA fingerprinting in forensics. It also highlighted its ability to both point to the guilty and exonerate the innocent.

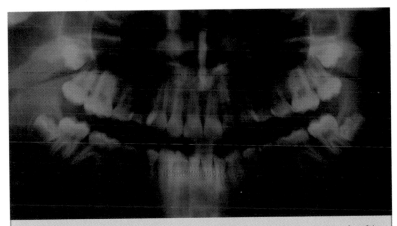

Inherited from the mother's side, mitochondrial DNA is used as an analytical tool in cases where old sample cells lack a nucleus, such as what happens in teeth over time.

MITOCHONDRIAL DNA

Not all DNA in cells is contained within the nucleus. Another form of DNA resides in mitochondria, structures outside the nucleus that produce energy that fuels cellular functions. Unlike nuclear DNA, which is inherited from both parents, mitochondrial DNA is inherited from only the mother. Mitochondrial DNA has taken on increasing importance as an analytical tool, especially in cases where the nuclei of cells have been damaged or destroyed, or in old samples in which cells lack a nucleus, such as bones and teeth (which normally have nuclear DNA, but may lose it with time). Testing with this sort of DNA can link an individual to a maternal bloodline but cannot distinguish between individuals in that line. Even so, testing of non-nuclear DNA has greatly helped determine identities, especially when maternal DNA samples are available for comparison.

How DNA Testing Works

DNA samples for testing are taken from body tissues, fluids, stains, and other biological evidence recovered at a crime scene. Blood placed in tubes, swabbed from a person, or absorbed from a dried stain can be used as testing material. So can hair follicles, though a hair without a root cannot—the hair itself does not contain DNA. Semen is a frequent source of DNA in rape cases. Urine and even urine stains are valuable sources of evidence, as is saliva, found in items such as cigarette butts and discarded wads of chewing gum. Samples from suspects are often in the form of oral swabs.

EXTRACTING DNA

In the laboratory, DNA is extracted from cells in the sample with enzymes that breach the cellular membrane. Enzymes cut the DNA into pieces, then the DNA is sorted out by length by passing them through a gel, made from gelatinous material in seaweed, that is subjected to an electrical current. Longer fragments contain specific areas of DNA belonging to the one-tenth of 1 percent of DNA that differs from one person to the next. These areas are used as markers, which will be used later for comparison between samples from the crime scene and those from the suspect or,

depending on the case, the victim. During the process, the attachments between rungs, or bases, in the ladder of these markers are severed, splitting the double DNA chains into two separate lengths. The DNA is then transferred to a nylon membrane, which serves rather like an examination table. Meanwhile, synthetic pieces of DNA are produced and fitted with radioactive tags, with the resultant product called a "probe." The DNA on the membrane is treated with a solution containing probes. The bases on the probes are attracted to and hook up with the corresponding bases on the marker sites of the sample DNA. The radioactively tagged markers show up as exposures on x-ray

film, enabling testers to identify them. The pattern on the film, which resembles the bar code on envelopes and packaging of retail items, is the DNA fingerprint. The presence of a single matched marker in samples from two individuals is not sufficient to make a positive identification. What determines the fingerprint is the number of times these markers are sequentially repeated at specific key sites. The repetitive sequences in DNA used for identification go by the acronym STR, for "short tandem repeat."

OTHER USES OF DNA TESTS

DNA testing for investigative purposes is not done only on humans. Animals are often the focus. While

Above: In a crime laboratory, DNA is extracted from the evidentiary samples and is subjected to a series of tests to determine if it matches a particular suspect in a crime. Top left: Bloodstains and other bodily fluids can be carefully extracted from a crime scene for DNA testing.

techniques may not be sufficiently advanced to identify individual animals, they can easily determine between one species and another. In itself, this use of DNA testing is a powerful asset in investigations of the illegal wildlife trade when it comes to determining that a hide or an animal product came from a protected species.

A new area of concentration is the testing of animal DNA by veterinary genetics laboratories to aid investigations of people-on-pcople crimes. A rape suspect pled guilty when dog urine on his truck matched that of the victim's pet, placing the rapist at the crime scene. The molester of a young boy went to jail after investigators matched the DNA of dog saliva found on the suspect with that of the boy's dog. The victim had described the place where the dog had licked his tormentor, and that is where police found the saliva.

Thus far, the site of DNA testing is in the laboratory because of the size of the equipment needed and the controlled conditions required to separate the DNA into testable strands.

Eventually, however, DNA testing at crime scenes may become routine. Researchers in the United Kingdom are working on a way to bring portable equipment and test DNA on the spot.

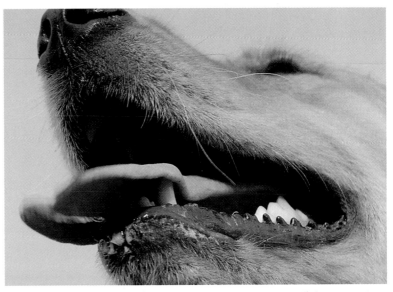

Often, saliva or urine samples taken from a victim's pet can be matched to traces of the same saliva or urine on a suspect, placing them at the scene of the crime.

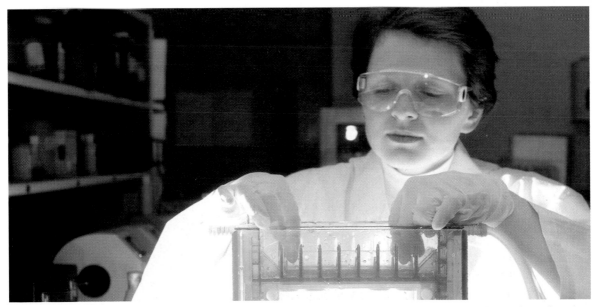

Currently, all DNA testing is done in a laboratory because it calls for sophisticated—and stationary—equipment. In the near future, however, DNA may be tested right at a crime scene. Researchers in the United Kingdom are now working to develop portable DNA testing equipment.

DNA Successes

There is no disputing that DNA testing has been a revolutionary advance in criminal investigation and has been highly successful in the effort to apprehend and convict criminals. At the same time, it is by no means the ultimate weapon against crime and is far from a guarantee of conviction. Absolute figures are difficult to come by, but a reasonable assessment of the conviction success rate achieved after investigations in which DNA matches have been made in the United States indicates fewer than half of the cases, perhaps as low as a quarter.

DNA IN THE UK

Despite its smaller population, the United Kingdom has solved many more crimes via DNA analysis than the United States has. Part of the reason is that Great Britain has what many recognize as the best DNA database for matching samples with suspects, due largely to a uniform national policy that dictates under what circumstances and from whom DNA can be drawn. British authorities have much more leeway than their U.S. counterparts and can even sample traffic offenders. In the United States, rules are much stricter and vary from federal regulations to those enforced by states within their

borders. Only a handful of states even collect samples from all convicted felons. The national DNA database, the Combined DNA Index System (CODIS), maintained by the FBI, contains DNA profiles of almost 3 million offenders but depends to a substantial degree on data collected by individual states.

Nevertheless, the number of cases solved largely through DNA matching continues to mount, especially in the area of "cold cases," law-enforcement jargon

for cases that have been open and unsolved for years. A typical case involved a man convicted 32 years after he committed a rape in New York City in 1973. The 58-year-old had been tried in 1974 but released on bail pending retrial after the jury failed to reach a verdict because his victim had not seen his face and could not identify him. While out on bail, he disappeared and did not surface until he tried to buy a shotgun in Georgia, which required a background check.

Above: With the assistance of DNA testing, many cases that were left unsolved for decades can be reopened and new, DNA-tested evidence can lead to a long awaited conviction. Top left: DNA analysis is more widely used in the United Kingdom than in the United States.

Since there was still a warrant out for his arrest, New York authorities were alerted to his whereabouts and resumed their investigation in order to bring him to trial once again. In the case file, they found underwear worn by the victim after the rape. DNA from semen discovered on the garment matched the DNA of the suspect's blood sample, taken decades earlier, leading to his conviction.

DNA matching also has solved many missing persons cases. A landmark Texas case occurred in 2004, when the first identification was made with the state's Missing Persons DNA Database, established by the state legislature in 2001.

Among the entries of the missing in the database was that of Donna Williamson, who had disappeared at age 19 in 1982. Skeletal remains believed to be hers were discovered in 1993, but positive identification could not be established. Her family submitted their DNA reference samples to authorities in 2003. Their samples were compared to the remains, and a familial connection was found. When coupled with dental records, the DNA evidence enabled authorities to positively identify the remains as those of the missing woman.

DNA matching is often used in missing person's cases, and a DNA Missing Person's Database was established in 2001.

A technician swabs a blood-stained dollar. Scientists can match DNA samples taken from objects left at a crime scene to a suspect's sample.

DNA Testing after the Fact

In January 2006, a cause célèbre of capital punishment opponents in the United States fizzled when DNA tests ordered by Virginia governor Mark R. Warner proved that a convict executed for rape and murder in 1992 was indeed guilty as charged. A report from the Centre of Forensic Sciences in Toronto, which tested DNA from Roger Keith Coleman, found that it matched, with a certainty of 19 million to 1, that of the assailant. This meant that there was only 1 chance in 19 million that someone else had murdered Coleman's sister-in-law, Wanda McCoy, in 1981.

Initial tests prior to the Coleman execution placed him within the 0.2 percent of the population that could have produced the semen found at the murder scene. The conclusion was challenged, first by his lawyers, and then by death-penalty opponents, including even the late Pope John Paul II. Shortly before the end of his term, Warner ordered the test due to an international outcry that mounted with the passage of time and the existence of testing procedures not available in the early 1990s. No executed convict had ever been exonerated by scientific testing before, but opponents of the death penalty

Above: In 2006, DNA tests proved that Roger Keith Coleman, who was executed in 1992 for the rape and murder of his sister-in-law, was in fact guilty of the crime. Testing that was unavailable at the time of the trial later helped prove that Coleman's claims of innocence were false. Top left: Blood samples can help prove if an inmate was wrongly incarcerated.

had high hopes that Coleman's protestations of innocence were true. Death-penalty proponents hailed the outcome.

CLEARING THE INNOCENT

DNA testing is a two-edged sword, however, when used after the fact of a guilty judgment. It has been instrumental in freeing some people who were wrongly convicted of crimes and imprisoned, sometimes for extended periods of time, and also clearing the names of those who had already served their time and been released. Additional independent DNA testing has been requested by prisoners and their attorneys, by state officials under pressure from prisoners' rights groups, and even by prosecutors who have found new evidence that contradicts a case's original verdict.

Five cases in point occurred in Virginia, ironically the same state in which DNA tests proved the guilt of the executed Roger Coleman. During 2005, five men convicted of rape during the 1980s and imprisoned were exonerated by DNA testing samples that had been saved by a conscientious staffer long after the relevant cases were apparently resolved. Two of the men had already completed their sentences. The other three were still in prison.

Testing was ordered after it was accidentally discovered that a forensic scientist who had worked in the state's crime lab from 1974 to 1988 had physically saved evidence samples such as blood smears and semen in her individual case files before they went into storage. With the chain of possession established and this significant advance in identification technology developed, random DNA testing of the files kept by the scientist, who died in 1999, was ordered and the men were cleared.

Some rights advocates worry, however, that appeals resulting in post-incarceration testing may be in danger. Some states—Florida and Ohio, for example—have passed laws eliminating the right of prisoners to seek DNA testing through the courts, restricting appeals only to prosecutors. Not all states, moreover, mandate that evidence be preserved throughout the time a prisoner is behind bars.

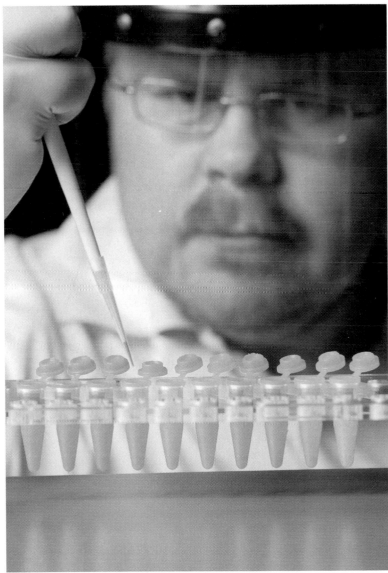

Left: DNA testing is often requested by prisoners and their defense attorneys who are seeking to reverse a conviction. Although this could conclusively prove innocence, some states have banned post-incarceration testing. Right: A technician analyzes samples in a laboratory.

Other Uses of DNA Testing

What do chardonnay, the prized white wine, and renowned reds such as pinot and gamay noir have in common? They all are cousins, of a sort. DNA testing at the University of California, Davis, has shown that 18 of the world's most prized grape varieties, or cultivars, are closely related, due to crossing of different strains. It may make a difference only to wine connoisseurs, but some of the grapes involved are not from among the aristocracy of vines.

MEDICAL USES FOR DNA

At a more compelling level, DNA profiling is being used to diagnose inherited diseases in fetuses and newborn infants. Among the disorders that may be so diagnosed are cystic fibrosis, hemophilia, Huntington's disease, Alzheimer's disease, and sickle cell anemia. Early diagnosis permits treatment to be planned well in advance and alerts medical practitioners and parents to the risks that lie ahead. Storage of DNA fingerprints of potential organ donors also has great promise. Matches between candidates for transplants and potential donors could be made much more swiftly with this resource.

DNA FOR AUTHENTICATION

DNA testing has, in fact, many uses besides its role in forensics. The insurance industry has touted this method as a means of deterring devious horse traders from bait-and-switch dealing. Occasionally, the purebred show horse that one pays for may not be the one that actually arrives at

Above: Not only used in forensics, DNA testing has recently shown that 18 of the world's most prized grape varieties are very closely related, but not all the grapes are from the best vines. Top left: DNA profiling can help test for inherited diseases such as cystic fibrosis.

the stable. Unscrupulous horse dealers have been known to switch horses from top bloodlines with those of inferior lineage once the deal has been made and payment given. Even a DNA test of semen sent with a horse may not guarantee its genetic worth. Semen samples, like the horses themselves, can be switched. The sure way to guarantee that the horse received is the one paid for is a DNA test of material taken from the animal itself, on the scene. That is enough, say experts, to catch the horse thief in the act.

The National Football League and the 2004 Summer Olympics have also used DNA technology, in these cases to ensure that legitimate memorabilia from these events would not be confused with fakes. Since the year 2000, footballs used for the Super Bowl have been marked with invisible, permanent strands of synthetic DNA, verifiable as to their identity virtually for the remainder of their existence by laser technology. A section of coded DNA from several unidentified Olympic athletes was mixed with ink marking all sanctioned goods from the summer games, certifying their authenticity.

POTENTIAL FOR FUTURE USE

The potential of DNA fingerprinting as a basis for banks of personal identification records is enormous. For example, the U.S. armed forces have begun to compile genetic records of military personnel, an identification measure that is far more effective than the traditional dog tags. The military hopes that DNA records, along with blood samples, will go a long way toward eliminating the phenomenon of the unknown soldier. Increasingly, there is a movement to have parents record the DNA records of their children as a way of finding them should they disappear. Given the grim connotations of such a program, some parents resist it. Civil libertarians also cast a watchful eye on broad programs of personal identification banks using DNA, due to implications about invading privacy.

To prevent show horse theft, DNA testing of the animal at the scene can help prove that the dealer sold the purebred horse he claimed to and did not switch it with an inferior animal.

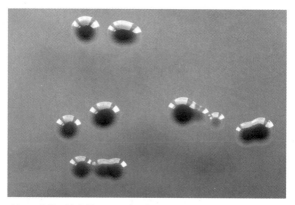

Microbes as Evidence

Microbes are humanity's curse and its blessing. Without the helpful microbes in our guts, we could not live, but microbes that spread disease have killed more people than all the wars ever fought. In an age when the threat of bioterrorism looms large, the coupling of infectious microbes with war has brought new focus on the nascent discipline of microbial forensics, which eventually might have as much promise as DNA evidence as a means of solving crime.

The increasing importance of microbial forensics was called to attention at the 229th national gathering of the American Chemical Society in March 2005 by Virginia Polytechnic Institute researcher Randall S. Murch, a former FBI veteran who created the Bureau's Hazardous Materials Response Unit in 1996. "It is imperative," he said, "to establish robust microbial forensic capabilities" that would allow authorities to track down those responsible for biological attacks and prosecute them, thereby perhaps deterring such events in the first place.

DEVELOPING PROTOCOLS

The National Bioforensics Analysis Center, created in the Department of Homeland Security and working with the FBI, reported on progress in the field to the White House during 2005. The FBI has created a committee of scientists and law-enforcement officials to develop methods of using microbiology as a forensics tool and has been working on the problem

with the American Academy of Microbiology. The techniques used in microbiological forensics are not especially new. Like humans, each type of microbe has its own genetic fingerprint. Databases of microbial DNA fingerprints are already being de-

veloped for organisms that cause airborne diseases and *Salmonella* infections. Tracing the sources of microbial disease outbreaks, such as those caused by the *E. coli* bacteria or the HIV virus, has long been a focus of the discipline of molecular epidemiology. Experts in the field become involved, for example, in tracking major outbreaks of food poisoning.

It is not so much the science that has to be developed but the methods and procedures that will result in evidence admission in court, as well as the tools that will prove useful to disease hunters in agencies such as the U.S. Centers for Disease Control. Prosecutors will have to prove, for example, that infectious agents found in the possession of a suspect are the same as those recovered at a crime scene. That is not an easy task, since many infectious microbes are present in the general environment. Microbial forensic techniques already have been used in a few cases, such as when people have deliberately tried to infect others with the HIV virus. A 1998 case in Louisiana resulted in a conviction later upheld by the U.S. Supreme Court.

Above: Microbial disease outbreaks, such as those caused by the E. coli *bacteria pictured here, have long been studied by molecular epidemiologists, and have been used to track the major outbreaks of the disease. Top left: Microbial forensics can help prove—and prevent—the link between bioterrorism and infectious disease.*

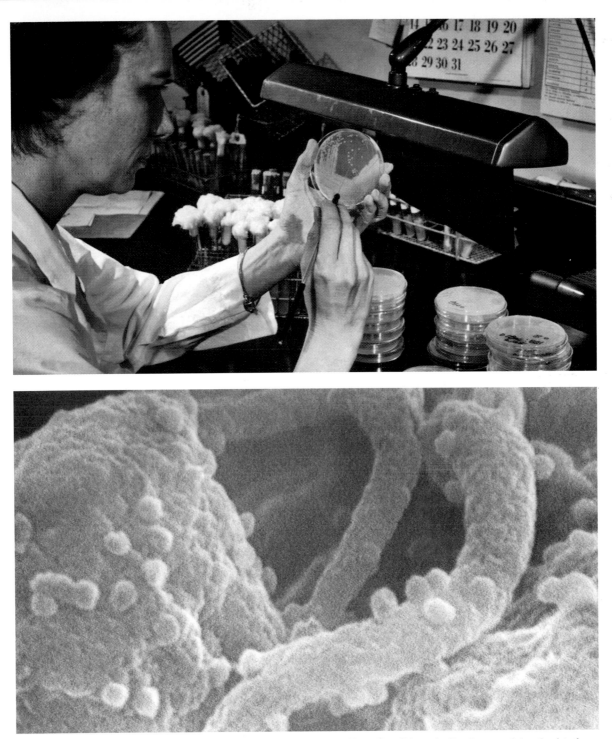

Top: A scientist at the Centers for Disease Control isolates Salmonella, Shigella, *or* E. coli *bacteria. Databases are being developed that will track DNA fingerprints for these organisms that cause airborne diseases. Bottom: The Human Immunodeficiency Virus (HIV), pictured here, is responsible for AIDS, and can be tracked by molecular epidemiologists in order to help prevent huge outbreaks.*

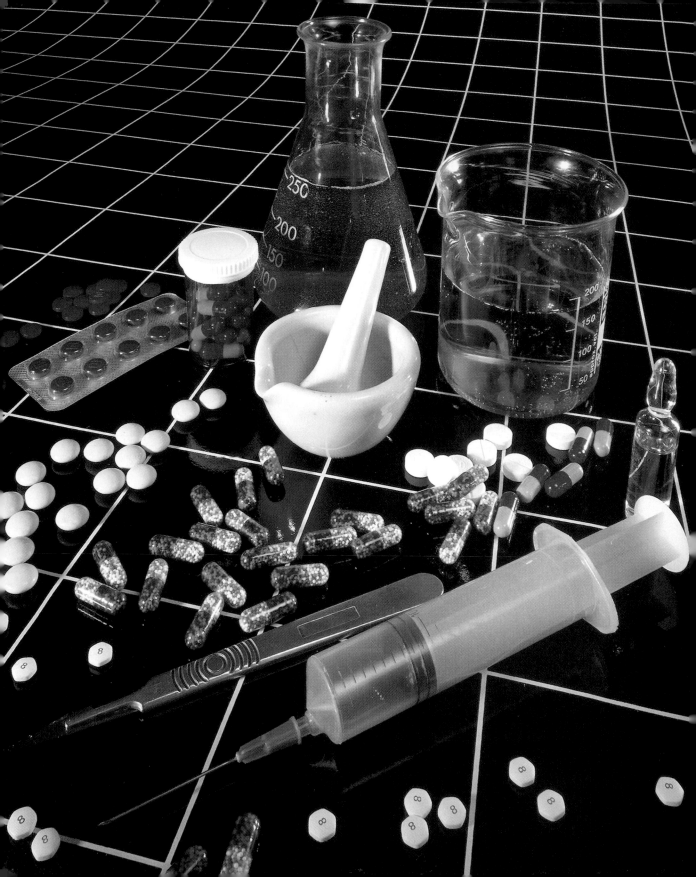

CHAPTER 8

THE CHEMISTRY OF CRIME

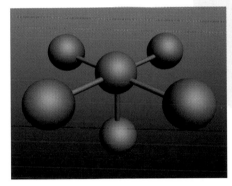

Left: The science of chemistry is an important forensics tool, crucial to the identification of illegal drugs and other chemical compounds, such as poisons, that may have been used in a homicide. Top: Forensic chemists identify the molecular structure of different types of physical evidence. Bottom: Trace evidence such as gunpowder residue is studied using chemical testing to determine the approximate distance between victim and shooter.

Chemistry is often described as the "mother of all sciences." Although some other scientists—physicists and astronomers, perhaps—may challenge this characterization, it has considerable support. Chemistry comes into play in virtually all the other sciences and their spheres of investigation. Ultimately, it examines matter in its most elemental form: the various atoms that make up the chemical elements. The change or rearrangement of just an atom or two in a molecule of matter makes that molecule—and therefore that matter—something entirely different from what it was, and chemistry's role is to track such changes.

Chemistry's broad reach touches upon myriad areas of forensic investigation. Indeed, some forensic laboratories have designated chemistry sections, which work in tandem with those involved in analyzing trace evidence, materials, and firearms. Whether or not a laboratory has a unit specifically assigned to chemical analysis, chemistry as a forensics tool is an essential component of several investigatory areas, especially toxicology, identification of illegal drugs and other chemical compounds, and arson.

Testing Techniques

Whether it is a poison suspected in a homicide, a suspicious powder found in a raid on a clandestine drug lab, or a liquid that may have been used to start a fire, these chemical compounds must be broken down so their components can be precisely identified and offered as evidence. Some such testing is done with techniques that would be familiar to almost anyone who has tinkered with test tubes in an analytical chemistry class—using reagents to crystallize compounds or to bring out trademark colors, for example. All drugs are acidic or basic, falling at either end of the pH scale, and chemical analysis allows forensic scientists to extract and identify drugs from evidentiary samples. Most truly significant testing, however, is accomplished with sophisticated instruments, used in a variety of forensic applications.

CHROMATOGRAPHY

Chromatography consists of tests that separate the different chemical components of a complex compound so that each may be identified. There are several methods of chromatography, but, essentially, each identifies components by the rate at which they migrate through a chromatographic device when passed

Confiscated items from a backpack—a powder, syringe, burnt spoon, and baggie filled with marijuana-like leaves—will be tested to confirm the presence of illicit substances.

through it in a carrier medium. The device may be a column or a coating on a plate.

Gas chromatography is one of the most commonly used methods of chromatography in crime labs. In this process, a test sample is vaporized and then injected into a pressurized cylinder—the column—with an inert gas, such as helium. Some columns are very long—up to about 30 yards—and have a skin about the thickness of heavy

Right: Suspicious powders and other compounds can be broken down into specific components for identification through chemical testing. Top left: Laboratory equipment ranges from the simplest test tube to sophisticated instruments such as mass spectrometers.

fishing line. As the vapor of the compound under analysis is carried through the gas, its components break out from the mixture. This breakdown is caused by the amount of each component that is absorbed by the gas. The greater a substance's tendency toward adsorption, or adhesion to the coating on the inside of the column, the slower it moves. Therefore, each component of the vaporized compound migrates at a different rate. On the vapor's arrival at the other end of the cylinder, these rates are registered by a detector, which electronically signals a recorder to create a graph showing each identified component.

Liquid chromatography works on quite a similar principal except that instead of gas, the carrier is, as the process's name implies, a liquid. Thin-layer chromatography, another method frequently used in forensics, uses a gel such as silica or cellulose to adsorb the components of the sample. It is the least expensive and fastest method of chemical analysis, but its results are not as definitive as those of other methods.

SPECTROSCOPY

Gas chromatography can be combined with another technique, mass spectroscopy, which identifies atoms and molecules by their respective masses. It works by ionizing the sample compound, and then sending the ions through an electric or magnetic field, or both. The paths of the ions bend according to the ions' masses, with ions

of lighter mass diverging most. The ions spread out in a spectrum defined by their differing masses, which is recorded on a photographic plate. Chemical compounds, even the most complex, have unique mass spectra; the precision with which mass spectroscopy has enabled scientists to identify various compounds has led scientists to compare mass spectra to fingerprints. Other types of spectroscopy utilize the electromagnetic spectrum to analyze the absorption, transmission, or reflection of wavelengths of light to identify the atoms of chemical elements.

Particularly in toxicology, mass spectroscopy, especially when combined with gas chromatography, is used in what are called confirmation tests. These tests confirm initial tests that

screen samples, often in large numbers, and cast a wide net for components they may contain. Confirmation testing zeroes in on components of particular interest and verifies their presence and identity.

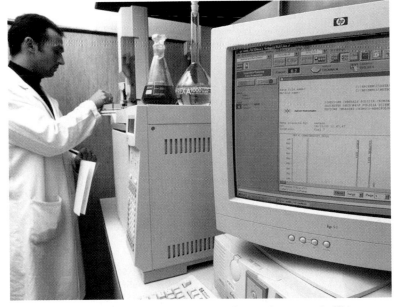

Above: A forensic scientist uses a gas chromatograph to analyze a sample. This instrument is used to separate mixtures of volatile gases or liquids into their various components. Top: The inert gas helium, which is commonly associated with blowing up balloons, is used in gas chromatography analysis.

The Role of the Toxicologist

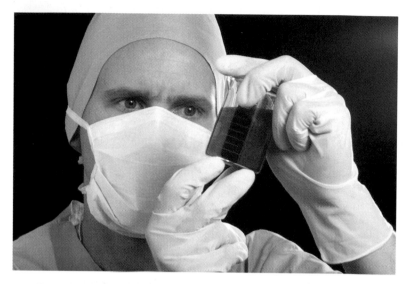

Toxicology, the science of the harmful effects of chemicals, including poisons and drugs, on living organisms is, along with pathology, one of the core disciplines of the forensic sciences. In fact, toxicologists often work in tandem with pathologists, the latter dissecting the body and the former chemically analyzing organs, blood, urine, stomach contents, and other bodily materials. Toxicologists are therefore called in on cases in which poisoning or drug overdose is suspected as the cause of death. They also may be involved when a victim has been clubbed, shot, or stabbed to death, in case drugs or poisons were also involved.

AN IMPORTANT ROLE

Basically a branch of analytical chemistry, toxicology often draws the interest of not only chemists but also people with a background in sciences such as biology and pharmacology. The part played by a toxicologist in a homicide investigation is critical when there is a need to determine whether drugs or other chemicals are involved and, if so, what role they played in the victim's death. The toxicologist analyzes fluid and tissue samples, and then helps interpret the findings. Here is where extensive

Top: Toxicologists are called in to determine the concentration, makeup, and effect on the body of illicit substances that may be involved in crimes. Bottom: When the use of so-called date-rape drugs is suspected, toxicologists can provide testimony about the effects of such substances. Top left: Various methods and instruments are used in forensic chemistry labs to analyze fluid and tissue samples.

knowledge of analytical chemistry comes into the picture. The image of a toxicologist peering into test tubes filled with multicolored liquids is somewhat off the mark. More likely, he or she will be operating sophisticated instruments used in gas and liquid chromatography, mass spectroscopy, and similar techniques.

The same approach comes into play when investigators are trying to determine whether drug or alcohol use impacted an accident or the commission of a crime, or when a defense counsel tries to employ drug or alcohol use as a mitigating factor during a trial. These inquiries often hinge on not only the presence of drugs, but also their concentration and effect on the individual when the event occurred. A toxicologist called upon to render an opinion in this kind of case needs considerable knowledge not only of the chemistry of drugs but also of clinical and medical studies into their effects on the human body.

SOCIETAL AWARENESS

Certain trends in society, including attitudes toward drug and alcohol use and emphasis on women's rights, have meant expanded roles for toxicologists. Toxicologists play a part in workplace drug and alcohol testing. As awareness of the use of date rape drugs has increased in the law enforcement establishment, toxicologists are called in to provide laboratory expertise and testimony when these substances are involved in crimes. Another field that may be broadly considered under forensic toxicology is the testing and evaluation of pharmaceuticals and other chemical products to determine their safety for marketing to the public.

TAKING THE STAND

Due to the scientific complexity of toxicological testing, toxicologists called on to explain their research and offer expert opinions in court must be proficient at explaining science in terms judges, attorneys, and, of course, juries can understand. Often, someone from the toxicology laboratory, either the person who has performed the analysis or a representative who can translate the findings into terms appreciated by the layperson, must appear in court. He or she must be able to explain not only the identity of a chemical and where it was found but often also the biochemical reactions that it triggers in the body of a victim. The toxic effects of poisons are sometimes due to the way the body metabolizes them, using enzymes, light, and oxygen. The metabolites thus formed may attack organs other than those in which the toxin resides; lead accumulates in the bone marrow, but its effects are lesions in the bone and soft tissue.

Toxicologists are often called to testify in court as expert witnesses.

Poisons

What makes a poison? The sixteenth-century Swiss alchemist and physician Paracelsus put both the question and the answer this way: "What is there that is not a poison; all things are poison and nothing without poison. Only the dose determines that a thing is not a poison."

LEVELS OF LETHALITY
A poison is a material that upon intake—ingested, breathed, or absorbed through the skin—can cause bodily damage or death, a definition that covers a vast amount of ground. Nevertheless, it is true, with the caveat expressed by Paracelsus. Virtually

anything you consume or inhale can kill you, but its lethality depends on how much and how fast you consume it. Even water can be a poison. If you drink gallon upon gallon of water a day, your potassium and sodium levels may be so diluted that coma and death can result. Rarely, people become so obsessed with the desire for water that this happens, but not overnight and not until they drink enough to drain a reservoir. Obviously, someone planning a homicide by poison would not use water. A better choice would be a compound called tetramine, which is used in some rat poisons. Five milligrams can kill a person.

The poisons of choice for homicidal purposes are those that are highly lethal, particularly in small doses, act quickly, and are difficult to detect. Also, poison can be an effective murder weapon if small amounts delivered over time go unnoticed until they build up in the victim's system to a lethal dose. For these reasons, human remains often go to toxicologists for examination even if it appears that something other than poison could have been the cause of death.

A DECEPTIVE KILLER
Obvious symptoms of poisoning can be the same as or similar to symptoms caused by other

Paracelsus (1493–1541), a Swiss physician and alchemist, is sometimes called the father of toxicology.

Above: The U.S. Food and Drug Administration is responsible for determining safe, nontoxic levels for fertilizers and insecticides used on food products. Top left: Even water can be poisonous if too much is ingested over a short period of time.

agents. Arsenic causes vomiting and severe diarrhea, but so do food poisoning and some types of influenza. Since poisons generally work at a cellular level, their presence often does not cause changes in the body that are apparent during an autopsy, especially if the poison involved was selected for its latent qualities. In the August 2004 issue of the *FBI Law Enforcement Bulletin*, FBI behavioral experts Arthur E. Westveer, John P. Jarvis, and Carl J. Jensen III noted, "All too often, authorities may certify a death as due to natural or unknown causes, resulting in important evidence of the crime being buried with the victim. Therefore, a great number of homicides by poisoning are detected only upon specific toxicological analyses carried out after the exhumation of the victim's remains."

The experts cited several cases to prove their point. One occurred in a small country town. A man died at a local hospital 10 days after being admitted for what his family claimed was pneumonia. Authorities cleared the body for burial and he was interred. They did not know that his wife, involved in an affair with another man, had collected $50,000 on her late husband's insurance policy and was pressing her lover to marry her. In a classic example of the stupidity so many criminals exhibit, she made an incomprehensible mistake. A suspicious fruit grower reported to police that she had returned to him a highly toxic herbicide. After investigating and finding out about the policy and her

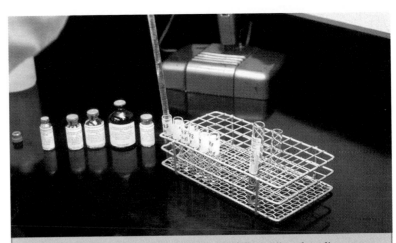

These samples are being tested for the presence of Clostridium botulinum, the bacteria responsible for the condition known as botulism. Toxins are poisons that are produced by living organisms.

DEFINING POISON

Strictly speaking, the words "poison" and "toxin" are not interchangeable, as many people may assume. In technical terms, a poison is a substance—more accurately, a compound—that, when taken into the body, damages health or destroys life itself. A toxin is a poison, usually containing proteins, that is produced by a living organism. The toxin that causes botulism, for example, is manufactured by a bacterium. Poisons can be the products of living things, but they can also be inorganic, including metallic elements such as arsenic.

Snake venom, according to the definition above, is therefore a toxin. So are skin secretions produced by the brightly colored arrow-poison frogs of the South American tropics. To obtain the poison to coat the tips of blowgun darts and arrows, native peoples may simply scrape the darts across a frog's back or, more gruesomely, may roast the frogs to obtain their secretions. Thus tipped, the points are lethal. Many zoologists distinguish between "venomous" animals, such as snakes, which inject venom, and "poisonous" animals, such as the arrow-poison frogs, which contain toxins in various parts of their bodies.

affair, police exhumed the body. The herbicide, paraquat, was found in the victim's body. The murderous wife was convicted and sentenced to prison and treatment in a mental hospital.

Poisons were weapons of choice for murderers from antiquity through the start of the modern era. Today, however, they have largely fallen out of favor. As a rule, poisonings account for less than 2 percent of the homicides committed in the United States each year. All told, about two million cases of poisonings of all types, including drug overdoses, are reported annually, with approximately 15,000 of them fatal.

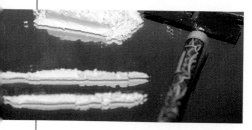

Drug Overdoses

The 38-year-old Oklahoma man was known to have a history of cocaine use. Eventually, as with so many other unfortunate addicts, there came a day when he overdid it. He was found lying on the bathroom floor, dead. When police arrived, it was not at all difficult to surmise what killed him. His arms were marked by three punctures from a bloody needle. Beside him, on the floor, was a spoon containing dried residue. It appeared that he had been injecting himself with powdered cocaine that had been dissolved in water. When injected, cocaine is released directly into the bloodstream in 15 to 30 seconds; the process takes three to five minutes when the drug is inhaled, or snorted.

TOXICOLOGY TESTING

Even with all the obvious evidence, the cause of the man's death was only a supposition until his remains were subjected to toxicological testing. Test results proved the obvious. In technical terms, he died from unintentional cocaine toxicity, in other words, a drug overdose.

Investigations of deaths by drug overdose and from alcohol poisoning constitute most of the work done by forensic toxicologists. Most, by far, are accidental, although, especially in the case of drugs, suicide may be involved. Overdoses are usually caused by using so much of a drug at one time that it is suddenly fatal, as opposed to long-term use of drugs in moderated quantities, which can ultimately lead to organ failure and death as well. An overdose can also occur when drugs are mixed—cocaine, amphetamines, and alcohol, for example.

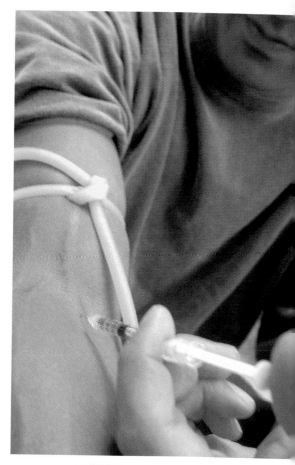

Right: Some users inject cocaine rather than inhaling it in order to get an almost immediate high, increasing their risk of overdosing. Below: A dried opium poppy seed pod. When these pods are still green, they contain the milky substance used to make heroin and other opiates. Top left: More deaths are caused by impurities in street heroin than by overdoses of the drug.

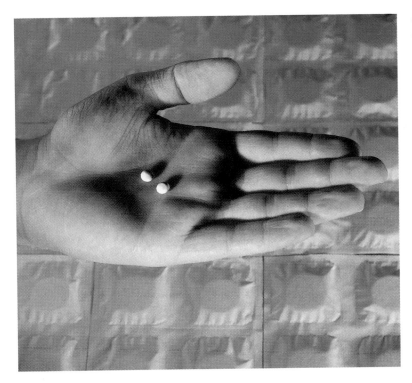

antagonists block the reception of heroin by the body. Perhaps most deaths from heroin use are caused by additives that adulterate street heroin or because the purity, and therefore the strength, of the heroin was greater than the user realized.

Overdose with so-called designer or club drugs, such as ecstasy (MDMA), and date-rape drugs, such as GHB or "scoop," is especially dangerous, because the body can be overcome very quickly. At high doses, ecstasy can cause the body temperature to increase rapidly, leading to muscle breakdown and failure of the kidneys and cardiovascular system. Fatal heart attacks, strokes, and seizures have been linked to the drug. When used to incapacitate women for nonconsensual sex, the date rape drug GHB is usually slipped into alcoholic drinks. Quick acting and fast dissipating, GHB is sometimes difficult to detect in emergency rooms. Effects of overdose include vomiting, loss of consciousness, impaired breathing, and often death. Its use in alcohol magnifies its depressant effects and can lead to respiratory depression, coma, unconsciousness, and overdose.

GHB is most commonly mixed with drinks such as flavored liqueurs, margaritas, and Long Island iced teas to cover its unpleasant, salty taste. This technique was used by three men who were successfully prosecuted after arrest by the Los Angeles County Sheriff's Department for drugging and raping 10 women and poisoning 6 others.

Above: MDMA, commonly known as ecstasy, is usually ingested in pill form. Left: GHB and other date-rape drugs that are often administered to victims in flavored alcoholic drinks can lead to impaired breathing, loss of consciousness, and even death.

HOW MUCH, HOW FAST?

The danger posed by an overdose depends on the drug. Despite what many people believe, heroin overdoses are not particularly common, nor do they occur quickly. They can also be immediately reversed with drugs called opioid antagonists.

Heroin, made from opium, is a member of a class of drugs called opioids, or opiates. Opioid

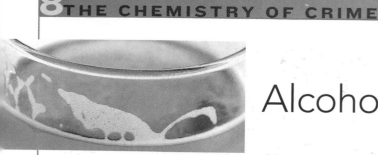

Alcohol

Homicides, diseases such as cirrhosis of the liver, and accidents in which drunk driving is involved aside, about 20,000 people die annually in the United States because of alcohol. The problem is not unique to Americans. It is worse in other countries, such as Russia, where the figure is almost double that of the United States.

Forensic toxicologists document the cause of deaths from alcohol poisoning just as they do those from overdoses of other drugs—by analyzing blood, organs, and other parts of the body with techniques such as gas chromatography and mass spectroscopy. When it comes to alcohol, however, toxicologists spend much more of their time on cases involving driving under the influence. A large forensics laboratory typically handles thousands of such cases annually, using techniques similar to those for testing fatalities related to alcohol.

FACTORS IN IMPAIRMENT

Even when a driver is not to the point of seeing two white lines where there is only one, the ability to safely handle a vehicle can be markedly impaired by alcohol, in this case, ethanol, or grain alcohol, the type used in beverages. Impairment can be influenced by many individual factors, such as a person's weight, whether or not the body is used to alcohol, and the amount of food consumed. As a rule, however, more than two typical alcoholic drinks in an hour can begin to impair motor skills and the judgment needed to operate a motor vehicle. Intoxication results from the inability of the liver to metabolize more than an ounce of alcohol an hour.

The amount of alcohol in the blood is an accurate reflection of the degree of intoxication. The level is based on the percentage of alcohol, expressed in grams, in the blood. Most states define the legal blood-alcohol limit for driving under the influence at

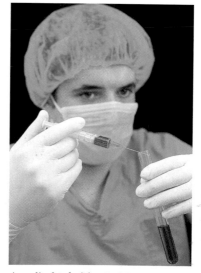

A medical technician prepares to examine a blood sample.

Above: Driving under the influence of alcohol is a much more common cause of death than alcohol poisoning. Top left: Alcohol impairment is influenced by various individual factors, such as the amount of food a person eats before or during alcohol consumption or the person's height and weight.

.08 percent, although there is a trend to lower that figure in many jurisdictions.

TESTING ALCOHOL LEVELS

Since blood testing must be performed in the laboratory, a person stopped for suspicion of drunk driving is generally asked to take a so-called breath test, usually by a Breathalyzer or similar device. Refusing a breath test may lead to various penalties, depending on prevailing law. Why a breath test? Because alcohol passes directly from the blood to the air sacs in the lungs, and the suspect's exhaled air provides an accurate indication of blood-alcohol level.

The definitive test, however, is performed on blood samples in the laboratory. These tests may or may not be done, however, depending on a host of factors, including legal wrangling. When drunk driving is suspected as the cause of an accident, though, a blood test usually is performed.

For accuracy, blood testing must be accomplished quickly, before the alcohol is metabolized by the suspect's body. If done in matter, it reflects the blood-alcohol level within a short time of when the sample was taken. The urine can also be tested for alcohol, but the urine test has major drawbacks for prosecution of a drunk-driving case. Alcohol concentrates in the urine for a day and often longer after intake; therefore urine samples may show a higher amount of alcohol in concentration and not reflect what was done within a few hours of the sampling.

Although many people think that alcohol is a stimulant due to the immediate effects, it is actually a depressant, which means it slows down body processes by diminishing the functions of the central nervous system.

WHEN CAN ALCOHOL KILL?

Alcohol is classified as a depressant. It inhibits the activity of nerves that control involuntary actions such as breathing, heartbeat, and the reflex that stops choking. Alcohol continues to move from the stomach through the bloodstream as long as the heart beats, even after the last drink is finished. Among the consequences of an alcohol overdose are choking on one's own vomit, cessation of breathing, a lowered body temperature that can lead to cardiac arrest, and a drop in blood sugar so precipitous that it causes a seizure. Death from alcohol poisoning can occur at a blood-alcohol level of .35 and higher.

Driving under the influence of alcohol is a serious offense. The legal blood-alcohol level in most of the United States is .08 percent, but varies throughout the world.

Clandestine Drug Labs

Forensic toxicologists deal with the detection, identification, and effects of drugs on and in the body. Many law enforcement agencies have forensic chemistry units whose job also is to analyze and identify illegal drugs, not in the body but at clandestine drug labs.

The Georgia Bureau of Investigation, for example, dispatches a Clandestine Laboratory Response Team of scientists drawn from around the state who help its officers crack down on laboratories where illegal drugs are produced. The North Carolina State Bureau of Investigation operates a Clandestine Laboratory Program that can draw on the services of more than 80 agents to respond when a suspected illegal laboratory is pinpointed.

The job of these teams includes assessing the safety precautions needed to examine

Right: Methamphetamine labs utilize whatever supplies are readily available, such as propane tanks as storage containers for anhydrous ammonia. Above right: A chemical technician in what is commonly referred to as a "hazmat suit." State and federal law enforcement officials who investigate clandestine labs wear these suits for protection. Top left: Agents from the Clandestine Laboratory Response Team employ many testing methods to confirm locations and sources of illegal drug labs.

and shut down secret laboratories, which are so volatile that they are designated as hazmat (hazardous material) sites. Because clandestine laboratories often are rife with environmental hazards, team members may wear protective equipment, such as a self-contained breathing apparatus for protection.

HAZARDOUS CONDITIONS

Among the dangers faced by raiders of illegal laboratories—besides the possibility of armed operators and guard dogs—is anhydrous ammonia, a liquid fertilizer sometimes used when ephedrine is reduced to make the drug methamphetamine.

The technique using anhydrous ammonia is known in the drug trade as the "Nazi method," because it is alleged that methamphetamines made this way were given to German soldiers during World War II to combat fatigue. Often the ammonia is

Methamphetamine labs are prone to explosions due to amateur handling of volatile substances.

stored in makeshift containers, such as propane tanks, creating a hazardous situation. Ammonia eats away at brass fittings such as those used on tanks, creating the potential for rupture. The ammonia produces suffocating, noxious fumes, which can severely burn the eyes, nose, throat, and skin and, if the exposure is significant enough, can be fatal. Sometimes, however, serious injury is averted as when three police officers in Iowa, without protective gear, suffered brief respiratory irritation from the chemical and were released from a hospital after treatment. Moreover, according to the federal Centers for Disease Control, anhydrous ammonia that is handled improperly can explode, with deadly results.

PROBLEMS IN CANADA

Canada has experienced a rash of home laboratories for manufacturing methamphetamine

Why are clandestine methamphetamine labs most prevalent in rural areas? Methamphetamine can be "cooked" using ingredients found in local stores, simple equipment such as hot plates and andryhous ammonia, which is readily available in agricultural areas where it is used as a fertilizer.

THE COUNTRY DRUG

Simple to produce, methamphetamine is sometimes called the "cowboy drug" or "country drug" because, unlike laboratories for most other illicit drugs, those that produce it are often in rural areas. The people who make methamphetamine do not need an urban setting; their equipment is as simple as hot plates, mason jars, and automatic coffee pots and their ingredients are as easy to obtain as ephedrine from dietary pills and cold medicines, camp-stove fuel, and iodine. Another factor that makes it easy to "cook" methamphetamine in the country is that anhydrous ammonia, because of its importance as a source of nitrogen for the soil, is commonly sold and stored in agricultural areas.

The price of anhydrous ammonia on the black market—as high as $400 a gallon—makes it extremely attractive to thieves. Even more tempted are the laboratory operators themselves, who steal the liquid from farms and agricultural supply shops. Occasionally, the theft backfires on the perpetrator. A man attempting to steal anhydrous ammonia from a storage facility in Washington State was severely burned when the valve of a 6,100-gallon storage tank was broken off.

and, especially in the eastern region of the country, club drugs. Canadian officials have warned that these laboratories pose a hazard not only to their operators—a man in Surrey, British Columbia, died when a methamphetamine lab exploded in his home—but to neighbors as well. If anhydrous ammonia is found at a laboratory, forensic chemists take samples for analysis. Their job entails identifying not only drugs recovered at the scene but also elements of the process used to produce them, additional evidence that is useful in court. Some testing occurs at the scene, but most is accomplished at a forensics facility, using methods similar to those that screen for drugs in the body.

Arson

A long with drug analysis, arson cases occupy much of the time forensic chemists spend in their laboratories. According to figures from the U.S. Fire Administration, a unit of the federal Department of Homeland Security, it can be estimated that more than 36,000 structure fires are intentionally set in a typical year, making arson the leading cause of fires. Arsonists torch about that number of vehicles annually as well. All told, moreover, the annual monetary damage to property totals a billion dollars or more. The human toll of arson, the second-leading cause of fire-related deaths, is more than 300 people.

WHY DO ARSONISTS START FIRES?

Arsonists are motivated by a host of reasons. Some arsonists are pathologically inspired to start fires. Fires may be set as a means of obtaining insurance payments, to cash in on policies for

Above right: About 30 percent of arson incidents in the United States are structure fires. Forensic specialists analyze crime scene materials for evidence of accelerants used in starting such fires. Below right: According to the U.S. Fire Administration, one in every four fire department responses is to a vehicle fire. Approximately 18 percent of these are due to arson. Top left: Abandoned buildings are targets for arsonists.

Gasoline is often used as an accelerant by arsonists.

buildings or automobiles. Some arson fires are ignited by vandals, one reason that arson seems to peak around Halloween, the Fourth of July, and New Year's Day. Some arsons are committed to cover other crimes, such as homicides. Arson is sometimes a means of extortion or revenge. When arson is used as a weapon, warn authorities, it can be especially deadly because it is specifically targeted to inflict personal harm. In 1999, for example, a revenge fire claimed the lives of six children in Minneapolis, Minnesota.

ANALYZING ARSON

A key job of arson investigation in the laboratory is the screening and analysis for accelerants of materials collected at the crime scene, as well as the identification of those accelerants. In technical terms used by law enforcement, this process is called fire debris analysis. Evidence recovery teams bring debris and materials that might be accelerants themselves back to the lab, usually in containers that are air tight to prevent volatile components from escaping. The container need not be anything more complex than a mason jar with a tightly sealed lid.

Forensic chemists often have to extract and concentrate small amounts of residue before testing them. Accelerant residue, for example, can be absorbed onto strips of activated charcoal, and then removed from the strip in a solution or solvent. The extracted sample is then tested with a technique such as gas chromatography. Gasoline is a typical accelerant, as is kerosene. Chemical analysis can easily tell the difference between the two in a sample. Chemists break the sample down, looking for the trademark components of the hydrocarbons in the fuel, such as benzene and naphthalene.

Once an accelerant is identified, it may be used not only as evidence of how the fire started but, depending on the other materials collected by investigators, to also implicate a suspect. For example, the accelerant found in fire debris may be matched with a sample from a suspect's clothing, indicating a direct link to the fire.

Although arson is a common crime, it is one of the most difficult to prosecute successfully. Only about 15 percent of arson cases lead to an arrest.

139

CHAPTER 9

PROFILING CRIMINALS

Left: Certain kinds of crimes, such as electronic embezzlement, may be motivated by greed, but the perpetrator may have additional incentives (such as debts due to gambling or drug addiction). Top: Although some people may consider drug use a victimless crime, a significant portion of crime can be attributed to drug users. Bottom: The average homeless person may seem threatening but is usually of harm only to him- or herself; a serial killer, on the other hand, is much more difficult to spot.

People are fascinated with the workings of their own minds. Yet, if there is anything about human mental processes that intrigues them even more, it is what goes on in the mind of the criminal. What makes a criminal commit crime? Indeed, what makes a criminal commit a particular category of crime? Some of these questions are not difficult to answer. Crooked corporate executives may cheat out of greed. An impoverished street person may grab a loaf of bread from a convenience store out of hunger. Anger and jealousy can fuel crimes of passion. Addiction can turn a user into a dealer in order to feed his or her drug habit.

Harder to fathom are the reasons some people commit particularly heinous crimes, those offenses that seem so far beyond normal human behavior that they are on the verge of incomprehensibility. Foremost among these are the atrocities committed by people who kill and sexually assault victim after victim, the serial killers and rapists. It is the need to understand the serial killer and the motives behind strings of repeated murders that has fostered the growth of criminal profiling as a forensic technique.

A Profile of Profiling

The term "forensic profiler" evokes a glamorous, somewhat mysterious image in the minds of many people. They visualize a person with an extraordinary and even psychic perception of the psychological machinations deep with the twisted criminal mind. This image may be cultivated by the entertainment media, even the news, but it is a phantasm. While some individuals are undoubtedly gifted in understanding the nuances of human behavior, investigative profiling—forensics profiling—is grounded in scientific analysis of the personality and motives of individuals who commit certain types of crimes and, often, specific criminal acts.

WHO ARE THE PROFILERS?

Profilers can be behavioral scientists, psychologists, or psychiatrists with a forensic background, or law enforcement officers with training in human behavior. They work on hunches, just as other types of criminal investigators do, and their insights stem not from any particular innate ability to read people's minds but, like those of forensic chemists and fingerprint technicians, from experience and knowledge of their scientific fields. At the same time, however, profiling differs from most other forensic

disciplines in that it does rely on the profiler's intuition, which is often based on experience. In this sense, forensic profiling is an art as well as a science.

MARGIN FOR ERROR

Given that human nature contains far more variables than, say, the number of electrons in an atom of a particular element, profiling can be relatively imprecise, and the profile itself may be strongly influenced by the interpretation of the profiler. Therefore, profiling's results may be more qualified than quantified, which can result in generalizations that, in the worst scenarios, can be stereotypical.

THE RISKS OF PROFILING

Stereotyping carries a notorious connotation in an age of political correctness, and profiling has created controversy when it has been attached to words such as "racial" and "religious." Unquestionably, innocent black and Hispanic motorists have been wrongly pulled over by police because of ill-founded profiling by skin color. Islamic dress and physical traits indicating Middle Eastern origins undeniably can trigger increased attentiveness on the part of airport security agents, especially during terrorism scares. Abuses occur

in the name of profiling, to be sure, but that does not mean that anything about a person, be it race, religion, gender, clothing preferences, age, and even work history, should not be factored into a legitimate forensic profile of a criminal personality.

PROFILING, AKA . . .

Even the name "forensic profiling" and, for that matter, the

Above: Arrest might occur because of a witnessed crime or because a suspect matches the description of an at-large criminal. Top left: Forensics profiling attempts to form an image or "profile" of certain kinds of individuals, who are prone to commit particular crimes.

activity it is meant to describe, can be represented differently and qualified depending on the situation in which it is used, its goals, and who performs it. Forensic profiling is also known as criminal profiling, criminal personality profiling, psychological criminal profiling, and offender profiling. Profiling can consist of an opinion that an adult who sets fires for pathological reasons is likely to have been physically or sexually abused, is a loner, and grew up in a dysfunctional family situation. It can be a list of characteristics believed to be common for people who hijack airliners. Or it can be an indication of where an unknown suspect, in law enforcement jargon an UNSUB, might live based on mapping and analysis of crime scenes within a given geographic area.

Profiling falls into two categories, based on whether it is done before or after a crime is committed. One attempts to profile the perpetrator of a crime that already has been committed by drawing a picture of the potential perpetrator from specific, known clues. The other, which creates a more general type of profile, involves development of personalities likely to commit particular crimes, such as serial killing and child molestation. The bulk of cases in which profiling is brought to bear are homicides; next in number are rapes, and often both are involved.

A Transportation Security Administration agent investigates suspect baggage at an airport security checkpoint.

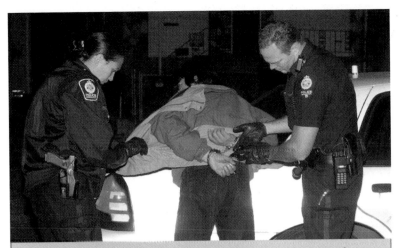

Profiling for certain types of crimes or criminal acts is lauded by some and denounced by others, with various studies bringing mixed results.

DEFINING PROFILING

Profiling has been described in many ways. It can be considered the evaluation of the full spectrum of evidence from a crime scene to hypothesize the personality of a perpetrator; the inference of distinctive personality characteristics of individuals who have committed crimes; the identification of a perpetrator based on the nature of the offense and the way in which it was committed; the psychological portrait of an UNSUB that points to the motives and kind of person involved; or an educated attempt to provide specific information about a certain type of suspect.

Law enforcement officials offer mixed reviews on the efficacy of profiling. There are those who view it as an important tool, while others are less enthusiastic about its success. A few studies have been done on the reliability of profiling. In an FBI survey of 192 cases in which profiling played a role, 88 cases were solved. Profiling proved helpful in identifying the suspect in 17 percent of the solved cases.

The Development of Profiling

Most experts believe that modern criminal profiling, as a forensics discipline, evolved from behavioral research by the FBI in the 1970s. The FBI made a concerted effort to apply principles of psychology and other behavioral sciences to the analysis of crime scenes. The primary focus of the effort was on serial killings, but it was expanded to cover many other violent crimes, including rape, child sex offenses, bombings, and bank robberies. Spurred by the FBI, many other law enforcement agencies began to use profiling in investigations when deemed appropriate.

THE ORIGINS OF PROFILING

The ideas that led to profiling, however, go back much further in history, to the early and middle nineteenth century. During the 1800s, there was a movement among some European scientists, particularly those engaged in anthropological theory, to determine whether a predilection for crime could be linked to certain physical characteristics. Out of this school of thought emerged a discipline known as criminal anthropology.

Inspired partly by the harsh theory of Social Darwinism, criminal anthropology was influenced largely by the writ-

Above: Acclaimed metaphysician Immanuel Kant (1724–1804) laid the groundwork for modern-day profiling. Top left: The details of a bank heist might assist investigators in predicting when and or where a future robbery might occur.

ings of German anthropologist and philosopher Jacob Fries and Italian surgeon and professor of medical law and psychiatry Cesare Lombroso. Fries, a disciple of Immanuel Kant, in 1820 was the first to suggest that the personality of a criminal and the nature of a crime could be linked, an idea that was a harbinger of modern-day profiling. Lombroso fostered the idea that science could define the nature of criminality in individuals, which also foreshadowed some of the concepts on which today's profiling is still based.

THE BORN CRIMINAL

Associated with criminal anthropology was the idea that criminality was inherited and could be reflected in certain physical characteristics. These features included a big, beaked nose, dark hair, big, fleshy lips, and a sloping forehead. Lombroso

Above: Early criminology had its missteps, including identifying the potential for criminal behavior by a suspect's "simian" physical characteristics. Below: The Nazi notion of racial purity, which doomed nonconforming individuals to imprisonment and death in concentration camps such as Auschwitz, proved to have no scientific credibility.

popularized the idea of the born criminal, identified by physical characteristics that were genetic defects, atavistic and similar to those of primitive hominids, apes, and monkeys. In the milieu of their times, such ideas about the inherited degeneracy of individual criminals were expanded by the proponents of Nordic racial superiority to indicate that entire races were genetically inferior. The Nazis utilized this line of thought to justify persecution of the Jews. Indeed, Fries has been accused of being an anti-Semite, evidenced by the fact that he associated with groups who claimed that Jewish influences were endangering the "purity" of German blood.

Although the concepts of criminal anthropology as originally defined are deemed themselves defective, the idea of using scientific principles to create a psychological portrait of criminality influenced early pioneers of criminology, such as Hans Gross. Credited with opening the door to the modern world of forensics in the late 1800s, Gross produced a ground-breaking book, *Criminal Psychology*, that explored criminality in terms of the human psyche.

PREDICTING BEHAVIOR

During the first half of the twentieth century, behavioral scientists in North America and Europe began dabbling increasingly in analyzing criminal behavior with an eye toward predicting it. The U.S. Office of Strategic Services commissioned a study of Adolf Hitler by a New York psychoanalyst; the study was so on target that it predicted he might commit suicide before surrendering to the Allies. A breakthrough occurred in the 1950s, when Dr. James Brussel, a psychiatrist with the New York State Commission for Mental Hygiene, accurately profiled Mad Bomber George Metesky and then Boston Strangler Albert DeSalvo. The FBI took notice, and the acceptance of profiling had arrived.

Profiling from Evidence

The same evidence that is used to re-create a crime scene may enable investigators to create an accurate profile of the criminal. Virtually any piece of evidence can be used as a piece in the profile puzzle: a brand of cigarette, an autopsy report, scratch marks on the breast of a murder victim, information from friends that a victim frequented a particular type of bar.

ANALYZING THE "MO"

This kind of profiling methodology, based on crime scene analysis, is sometimes called reactive, or, by some experts, deductive, profiling. The profile is created from analysis of all items of forensic evidence plus the overall characteristics of the scene—what may have been taken away and how entry was achieved, for example. Entry gained by jimmying open a window or picking a door lock provides clues to the *modus operandi* (MO), the manner of operation with which the offender commits crimes.

SIGNATURE BEHAVIOR

Unlike the MO, which in a sense is a trademark of the criminal, signature behavior is not an action needed to commit the crime but, in effect, a personal statement, such as making off with an item of the victim's clothing or one kind of valuable to the exclusion of others. Coupled with inferences deduced from the scene, an analysis of the victim's characteristics, called victimology, helps in creating the profile. Victimology is a thorough examination of a victim, ranging from lifestyle to hairstyle, which may provide clues to

Left: The kinds of items stolen, the time the robbery is committed, or the criminal's modus operandi may indicate a pattern to trained investigators. Above: Although carefully planned by an "organized offender," a car-jacking can go dangerously wrong during a high-speed chase. Top left: The brand of cigarette smoked or a DNA trace off a used cigarette butt can aid in the identification of a suspect.

why that individual was selected and, in turn, the psychological makeup of the offender. To put it simply, if a criminal butchers prostitutes à la Jack the Ripper, he may be sexually dysfunctional. If a murderer prefers blondes, negative interaction with someone who is blonde may lurk in his or her background.

TYPES OF OFFENDERS

Profilers have formulated systems for classifying types of offenders by analyzing crime scene evidence, but no particular method is universally accepted. The system used most, based on a model originally applied to serial killers that was developed by the FBI, lacks absolute scientific authority. It groups offenders into three classes: the organized offender, the disorganized offender, and the mixed offender. Some profiling authorities consider these terms an oversimplification and suggest that the terms "organized" and "disorganized" might be better replaced by terms used in psychiatric medicine: psychopathic and psychotic, respectively.

The reasoning behind the terms is that a psychopath understands right and wrong but does not care about the distinction, probably because of an antisocial personality disorder, while the psychotic is diagnostically mentally ill. A scrupulously planned crime, obviously premeditated, is the mark of the organized offender. This kind of criminal attempts to obscure evidence and perhaps to disguise the nature of the crime, for example, by setting a fire to conceal evidence of a murder. The body or a weapon may be removed and hidden somewhere far from the scene of the offense. If the disorganized offender has a plan, it is forgotten just before and during the commission of the crime, and the results are a chaotic mess, with key evidence, such as footprints, tools, or weapons, often strewn about and plainly visible.

A problem for profilers is that many, perhaps most, crime scenes do not neatly fit into the organized or disorganized category. In these instances, the perpetrator is classified as a "mixed" offender, one who plans the crime but loses control during commission of it. Moreover, there are few constants. An otherwise organized individual who is high on drugs, drunk, or in a monumental rage may leave a scene that can be described only as disorganized.

Above: An imprint of a shoe tread in soil or a grain of sand from a specific beach found in the crevasses of a boot can link the scene of a crime to a particular suspect. Top: The position of spent bullet casings can lead investigators to arrive at a clearer idea of what occurred at a crime scene, and who may have perpetrated it.

Profiling Projections

As criminal profiling began to take hold among law enforcement agencies during the 1970s, the federal Drug Enforcement Administration (DEA) developed profiles that agents and behaviorists believed fit couriers of illegal drugs. This was a major step in the system of proactive, or inductive, profiling, based not on evidence from a particular crime but on anecdotal law enforcement experience and both statistical studies of and generalizations about crime and criminals. This methodology is intended to profile the sort of person likely to be involved in a particular form of crime, a projection, in other words. This form of profile can be as simple as a policeman's intuition that a young man strolling the streets of a retirement community at midnight is up to no good or as complex as an assessment of why certain individuals are inclined to commit multiple murders.

SUBJECTIVE PROFILING

On the street level, proactive profiling may be based on factors that seem highly subjective on the part of law enforcement. Dress, grooming, being out of place in a particular situation, gender, and race may figure into the mix, often raising the red flag of controversy. Indeed, especially when it comes to race, abuses do exist. At the same time, consideration of race can be critical. For example, a profile of someone likely to be involved in a white supremacy organization would obviously exclude African Americans, and one would not expect that Caucasian racial traits would be shared by members of militant black groups.

Above: It is possible to predict criminal behavior from a study of demographics; for example, more than nine times the number of males are in prison than females. Above right: Certain types of crimes, such as hate crimes, are often committed by members of racially exclusive groups such as white supremacists. Top left: A vagrant loitering would be construed differently from a businessman waiting for a bus.

The Detroit Metropolitan Airport in Michigan initiated a program in 1974 to profile the actions and origins of drug traffickers; common suspicious characteristics cited were point of origin, signs of nervousness, lack of speed in deplaning, and flying with minimal or no luggage. Later models were adapted to suit various other airports.

STUDYING AIRPORT TRAFFICKERS

Among the building blocks of proactive profiling are clinical and informal studies of prison populations; personal experience of the profiler and associates; and data from a variety of sources, including demographic statistics and even magazine and newspaper articles, which are regularly surveyed by many law enforcement agencies.

The DEA drug courier profile began as an experiment at the Detroit Metropolitan Airport in 1974, during which agents observed airline passengers and gathered data on their appearance and behavior. Additionally, agents built in information about the known tactics used by drug traffickers and couriers across the country. Eventually, these profile models were adapted to many different airports.

A case often cited in the literature of jurisprudence involved a woman stopped at the Detroit airport who had been a passenger on a flight from Los Angeles. Much of the heroin on the streets of Detroit, agents believed, originated in the city of Los Angeles. Agents testified that the suspect was the last passenger to depart the plane and seemed very nervous, carefully surveying the arrivals area before entering it. She walked past the baggage claim without retrieving any luggage, and then changed the airline for her planned flight departing Detroit. Agents deemed that these facts were at least partly enough justification to confront the woman.

Introducing the Serial Killer

There are times when headlines in the press and television news reports can make one feel that serial killers lurk virtually everywhere, ready to strike with sadism and savagery anytime, any place. Dennis Rader, the BTK killer; David Berkowitz, Son of Sam; Jeffrey Dahmer, cannibal and necrophiliac; Albert DeSalvo, the Boston Strangler; Ted Bundy, torturer and rapist: One after another, as their names are attached to their repetitive homicides and they are apprehended, their images sear the imagination of the public.

IS SERIAL KILLING ON THE RISE?

Indeed, some authorities suggest that the rate of serial murders is increasing, although the seeming prevalence of these crimes may be due to increased press coverage and the enhanced ability of law enforcement, engendered partly by forensics, to solve them. In fact, science has not discovered a way to quantify serial killings, so an exact determination of the number of serial murders is not possible. As for forensics, profiling is tailor-made for tracking down serial killers, even when cases seemingly have gone cold.

A serial killer, by most definitions, is someone who murders again and again, usually one victim at a time, over a lengthy period, vanishing

Left: Dennis Rader, a sanitation inspector who became known as the BTK killer, shown here during his sentencing, blended in so well with his surroundings that it took a long time for investigators to link him to his many insidious crimes. Right: Although he confessed to murdering 13 women in Boston, Albert DeSalvo was killed in prison before ever coming to trial. Top left: Are there greater numbers of serial killers among us today than ever before? Some authorities believe the rate of serial murders is increasing.

into society during cooling-off periods between attacks. The FBI considers an individual who has killed at least three times before "cooling off"—the term is actually used by law enforcement officials—for a matter of days to years, a serial killer. Although many serial killings are considered sex crimes—sex of some sort plays a role in most—the underlying motives are not always obvious, being hidden deep in the murderer's mind or in his or her past.

Credit for coining the term "serial killer," which came into vogue during the 1970s, is usually given to either former FBI agent Robert Ressler or criminal investigator Dr. Robert D. Keppel, both famed for tracking down notorious serial murderers. British author John Brophy is also given laurels by some for originating the label.

NEW NAME FOR AN OLD CRIME

While the term may be of relatively recent invention, serial killing is nothing new. Records of multiple murders by one individual over time go back at least to the medieval period. Wealthy French aristocrat Giles de Rais is reputed to have kidnapped, raped, and killed more than 100 boys during the fifteenth century. A Hungarian blue-blood, Elizabeth Bathory—women serial killers are rare—is alleged to have killed at least 600 girls at the turn of the fifteenth and sixteenth centuries. So incomprehensible were their

Postal workers are not more likely to kill their coworkers than workers in other fields. Several high-profile incidents in postal facilities, however, have led the public to connect spree killing to this particular occupation.

MULTIPLE MURDERERS

The serial killer is not the only kind of murderer who kills large numbers of people. Forensic behaviorists recognize two other groups of multiple murderers. Mass murderers are those who kill more than four people in a single instance. This type of criminal has been described as "going postal," a phrase that originated in the mid-1980s. It was during those years that a number of disturbed post office employees or former employees killed several of their coworkers in sudden assaults with firearms. Spree killers, another group of multiple murderers, kill several people at different times and different places in a short period of time, without the cooling-off period typical of serial killers.

crimes that De Rais and Bathory were thought by the people of their time to be werewolves or vampires. Then, of course, there is the most famed serial killer of them all, Jack the Ripper, the terror of Victorian London.

THE "TYPICAL" KILLER

A number of overall profiles have been offered for the typical serial killer, the simplest being that he is a white male in his twenties or thirties who targets strangers near his home or workplace. A more detailed profile comes from leading criminologist and psychologist Dr. Eric W. Hickey from demographics he has compiled on serial murders. Eighty-eight percent of serial killers are male, and 85 percent are Caucasian; their average age when they commit their first murder is approximately 28.5 years.

Varieties of Serial Killer

Scientists have probed, with mixed results, the hellish recesses of the serial killer's mind for the motives behind the horrendous crimes these murderers commit. They generally agree, however, that sexual lust is the most common motive. At the same time, however, the cycle of cooling-off periods and killings do not seem well correlated with cyclic peaks and valleys in the killers' sex drives. Serial killings motivated by lust usually involve the most deviant and bizarre behavior of any sort. Cannibalism, necrophilia, sexual torture, and, more often than not, rape go hand in hand with murder. The sexual motivation for serial killing may be almost exclusively male because it seldom seems to play a part in the very few of these crimes committed by women, a large number of whom are so-called black widows, women who eliminate husbands one after another for financial gain.

GROUPING BY MOTIVE

Among the models for classifying serial killers is one that groups them by their supposed motives, rather than whether a crime scene is organized or disorganized. The visionary killer is one who is motivated to kill by delusions, voices in the head, or other hallucinations. The power- or control-assertive killer, believed to be the most common type, gains a sense of dominance, sexual and otherwise, over his victims. If he rapes, it is not for sexual pleasure but for a sense of inflicting his dominance over another. Sometimes murder is preceded by punishment rituals. The hedonistic killer, another sort, does kill, and often rapes or commits another sexual act

Above: Although the "black widow" or "femme fatale" killer is featured in numerous crime dramas, the vast majority of serial killers are men. Top left: A serial killer may not be lurking around every dark corner, but it is best to avoid isolated, poorly lit areas.

Even the most seemingly incidental item left at a crime scene might prove to be an important forensic link to a criminal. Telltale human hair, thread samples, or boot tracks can help investigators find a link to similar crimes and eliminate suspects.

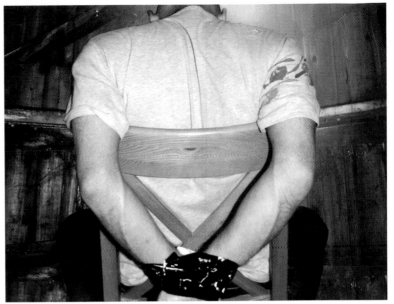

Left: Because serial killing is typically the province of sexually deviant males, the majority of their victims are women. Above: Some researchers postulate that serial killers engage in the bondage and torture of their victims in an attempt to gain a modicum of power and control that was lacking in their childhoods.

upon the victim, because he gets a thrill from it. Cases of this kind may involve cannibalism and torture. In cases where sexual gratification is not involved, the offender's pleasure can come from the thrill of the chase and the kill, except that the victim is a human rather than an animal. Still other serial killers, according to the model, are mission oriented. They are literally on a mission to eliminate a certain type of person, who they believe is in some way harmful.

PROBLEMS WITH MODELS

Most models of serial killers are not cut-and-dried. Murders can involve more than one motive, such as being on a perceived mission and hearing voices that order its accomplishment. Likewise, assessments of char-

acteristics shared by all serial killers are generalities but do seem to hold, at least for many of them. The forces that lead a person to serial killing probably start early in childhood and include a chaotic family background and physical and psychological abuse. Potential serial killers frequently build fantasy lives while quite young, often dwelling on unreality so heavily that it becomes, for them, a real existence. This fantasy life is a refuge, a protection from the unpleasantness of the real world, and may seem harmless at first but can drift into themes laden with violence. Whatever the subject of early fantasies, later ones often evolve into plays of rape and murder, which eventually may be acted out in real life.

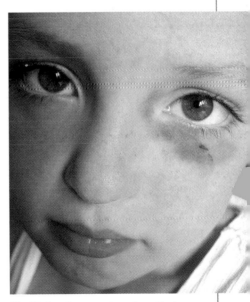

Familial maltreatment can be cyclical, with each generation inflicting pain on the next; psychological and physical abuse may contribute to future anti-social behavior, although the majority of those who suffered violent upbringings do not become serial killers.

Profiles of Arsonists

The fascination people have with fire may be the reason some arsonists share certain traits with serial killers. The National Center for the Analysis of Violent Crime, a unit of the FBI, uses the same terms for arsonists who set multiple fires as it does for multiple murderers: serial fire setter, mass fire setter, and spree fire setter.

REPEAT OFFENDERS

According to this classification system, the serial arsonist sets at least three separate fires over a lengthy period, with a cooling-off period between fires. The time span may last from a matter of days to several years. Spree arson occurs when three fires are set at different places, one right after the other. Mass arsonists set several fires at the same place within a short period.

Serial arsonists, who apparently derive a feeling of dominance from setting a blaze, are the most dangerous of the fire-setting trio. They strike randomly, in unpredictable fashion, and their motives—unlike those of arsonists who strike for profit or revenge—are shrouded.

Profilers note that this type of arsonist, like the serial killer, is typically a young male. He is usually a loner, often burdened by a feeling of inadequacy, and more likely than not poorly educated. Often, he is a poor planner and his preparations for setting a fire are hit or miss. Usually, the arsonist lives near the crime scene, sometimes so near that he will walk to it.

BACKGROUND OF AN ARSONIST

The history of serial arsonist arrests indicates that these criminals often had several brushes with the law before turning to setting fires. These incidents are usually minor and frequently involve substance abuse.

The serial fire setter is often involved in what profilers call "excitement-motivated arson," in which starting a blaze provides the satisfaction. In other words, the arsonist gets a thrill out of the process.

Above: Women rarely set criminal fires. Whether they are serial, spree, or mass arsonists, statistically, most arsonists are young men. Top left: Striking a match is just the beginning for the arsonist; a full conflagration is much preferred—and the ultimate goal.

Although an excitement-motivated arsonist does not mean to hurt anybody, things can get out of hand and bystanders or fire fighters may be injured or killed.

Setting fires can serve as an expression of anger or frustration. "Prank" fires set by young people may indicate more deeply serious problems that will lead to adult arson.

The sexual element in fires set for excitement sometimes is manifested by human waste and ejaculate found at the fire scene. Unlike serial killers, the excitement-motivated arsonist is not out to hurt people; he is instead out just to see something burn. Setting a fire can be the arsonist's call for attention.

Typically, the excitement fire-setter is a poorly educated young male. Lack of education, however, may not be peculiar to this type of arsonist and may be shared by arsonists in general. Studies show that most excitement fire-setters have very low IQs. Many, if not most, are single and live under the parental roof. Problems associating with the opposite sex and a feeling of social inadequacy also fuel excitement arsonists. Childhood abuse, of various types, also seems typical, as is an abiding anger and a need to strike out. Unhealthy anger, some psychologists believe, lies at the root of many fires set by youngsters that seem to be pranks but really are a sign of more serious and deeply rooted problems. If untreated, children of this type can develop into full-fledged adult arsonists.

Left: Dry conditions can cause a small excitement-motivated fire to turn into a raging inferno. Above: Playing with matches or lighting fires can be a disturbed person's call for attention.

CHAPTER 10

CLUES FROM NATURE

Left: Weather conditions can help forensic scientists to ascertain a time frame for an occurrence. Top: The changing of the seasons will assist investigators in determining the time of an event, such as a murder. Bottom: The presence of insect larvae or the remnants of a bloom may pinpoint the age of a body.

Storm clouds gather, blackening a sky split by flashes of lightning. The endless sneezing and sniffling of allergy sufferers signals that pollen is riding on the air. Crimson and gold leaves are falling, blanketing the woodland floor. Winds sweeping between red rock mesas send dust devils swirling across the desert floors. The end of winter's cold brings out winged insects to feed and procreate.

Natural events such as these are signs of oncoming weather, changes of seasons, and other normal fluctuations, large and small, in the world of nature. The scientific study of nature can be pursued for many reasons, including expanding our knowledge of our world and ourselves, as well as for pure personal enjoyment.

Some natural sciences, such as entomology, the study of insects, have been used to help solve crimes since the Middle Ages. Other disciplines, including meteorology and geology, have only recently become part of the techniques used in forensic investigations.

Forensic Botany

For many of those in the baby-boom generation, the phrase "flower power" may stir memories of hippies with scraggly beards, index and middle fingers spread in the V of the peace sign, and girls with long tresses handing out blossoms to soldiers bound for Vietnam. Today, forensics has a flower power of sorts as well. Botany and certain allied fields are important forensic sciences, sometimes providing clues about crimes that are available through no other fields of study.

PLANT PARTICLES

If a murderer drags a body through the woods or a burglar steps in a flower bed before prying open a window, traces from the plant world may cling to clothing, shoes, or hands and, if matched with plant types at the scene, make a persuasive argument tying place and person together. The murder of a woman in Minnesota was solved after a botanist examined the suspect's dew-soaked pants and found tiny particles from a branch that was growing near the victim's home. In Florida, bits of crushed plants led to the arrest of a man who was eventually convicted of dragging a teenage girl into the woods, throwing her down on a blanket, and raping her. A blanket with bits of weeds, leaves, and seeds was found in the car of the suspect, who claimed it had been used for a family picnic in a nearby park. The alibi failed when botanists found that the mixture of plant species

Left: The botanic residue found on shoes can link a suspect to a particular place, such as a flower bed outside a burglary victim's pried-open window. Right: Aquatic plants are specific to watery venues, and can provide clues to a victim's or a suspect's movements and even tie them to a particular body of water. Top left: Forensic botany is the study of plant life, applied to forensics.

set upon by a gang of youths who beat them with baseball bats, bound them with duct tape, and threw them into a pond. After one boy managed to free himself, he saved his friend. The pair went to the authorities, who apprehended three suspects. Police took samples of mud from the sneakers of both the victims and the suspects. Analysis showed that 14 genera of diatoms—algae encrusted with siliceous bodies—were present in all samples. So was an especially rare algae of another species. The species composition of algae from the samples was like that found in mud around the pond, evidence leading to the conviction of the attackers.

BOTANICAL DNA

DNA fingerprinting of plants can be a powerful tool for forensic plant scientists. Investigators with the Indiana Department of Natural Resources were seeking a tree thief who cut and removed a prized—and very valuable— walnut tree from a landowner's property. A molecular geneticist at Purdue University matched wood from the stump to a walnut log confiscated from a sawmill. Authorities tracked the log to a logger who had worked on an adjacent piece of property and had poached the walnut, which can be used for fine furniture and is difficult to obtain.

represented by the litter on the blanket corresponded with the place where the girl said she had been raped but not the park locale where the suspect allegedly had picnicked.

AQUATIC EVIDENCE

Knowledge of aquatic plants and related organisms often helps police investigations of crimes in or near bodies of water. Two young boys in Connecticut were

Forensic Palynology

Pollen, as people who suffer from allergies know, seems to be just about everywhere. When pollen is abundant, it is wafted on the wind, coats quiet ponds with yellow-green scum, and makes windshields look as though they have been in a dust storm. Someone untutored in the nuances of how pollen is produced and dispersed might think that the invisible clouds of pollen grains that plants dispense are a jumbled hodgepodge, mixed beyond comprehension. This is far from the truth. Scientists who study pollen, palynologists, can match pollens to their producing plants using the infinite variety of shapes and external configurations of the grains. This fact, and because pollen is all around us, has made pollen analysis a budding new field of forensics. In addition to pollen, palynologists analyze spores, the single-celled reproductive bodies of nonflowering plants such as fungi and ferns.

A BUDDING SCIENCE

Forensic palynology is a relatively new field, generally believed to have begun in Europe during the 1960s. One of the first cases in which pollen analysis helped solve a crime occurred in Austria. Police were investigating the disappearance of a male

Right: The spores of a mushroom plant can be analyzed in much the same way as pollen can. Above: Pollen travels in the wind to sometimes land on water, as seen here; different locales have different pollen mixes. Top left: Pollen can be identified by its specific shape and grain configuration.

Rather than leaving a clue at the scene, suspects can take evidence with them, such as mud in the tread of a shoe or boot or inside a pants cuff.

Pollen analysis lead Austrian police to the Danube River, pictured here, where a murder suspect had buried his victim on its bank near the site of an ancient hickory tree.

vacationer near Vienna. They had targeted a likely suspect but there was a major problem: no body. Police, however, did have something they considered an important clue: mud on the suspect's boots. A geologist who was a pollen expert was called in on the case. In the mud, he discovered pollen from spruce, willow, and alder trees, as well as from an ancient hickory, millions of years old. That combination of pollens, it turned out, was found in soil in one small area near the Danube River. Confronted, the suspect confessed and led police to the buried body.

Palynologists can trace samples of pollen taken from evidence—in the hair of a criminal suspect, for example—to the place where it originated, because different geographical areas have different pollen prints, according to what plants live there. The print can be determined from knowing the particular mix of pollen in the air in a given place. Many palynologists are also geologists. Pollen is an exceedingly tough material, highly resistant to decomposition, and is often found in geological strata millions of years old. Geologists regularly use it to date sediments during exploration for oil. Archaeologists also use pollen to date ancient artifacts and human habitation sites.

If any nation can be credited with pioneering forensic palynology, it is New Zealand. A classic investigation there was of the rape and murder of a woman jogger in a remote area of a city park. The crime scene had a relatively specific plant cover, mainly pine and fir. When police searched the apartment of a suspect who had been seen in the park, they retrieved dirty clothes from his hamper that were laden with pine pollen and fern spores. The combination was similar to that in soils at the murder scene as well as on the shorts of the victim. The amount of pollen on the clothing, moreover, indicated that the attacker and the victim rolled around on the ground. No area near the scene had pines and ferns growing together. The palynological evidence helped lead to the conviction of the suspect.

LEGAL LIMITATIONS

Despite its potential, the use of forensic palynology is not widespread among the world's law enforcement agencies. Not many pollen analysts are trained in forensics, and some who are, including Dr. Vaughn M. Bryant Jr., director of the palynology laboratory at Texas A&M University, one of the few centers of forensic palynology in the United States, say that many seem unwilling to handle the legal repercussions of working in forensics, such as cross-examinations in court and the potential consequences of testifying against a defendant.

The Properties of Pollen

The principle of evidence transfer—that anyone at a crime scene will leave a trace behind and pick up something from the scene—makes pollen perfect as forensic evidence. Pollen grains, which usually have rough outer skins, easily cling to surfaces. Thus, they attach to hair, clothing, skin, shoes, automobiles, and myriad other items that can be recovered from a crime scene, victim, or suspect.

ABOUT GRAINS

Pollen is a powdery mixture of grains produced by the male organs of plants—anthers in flowering plants, cones in conifers. The grains, usually spheres, are often covered with projections, like tiny golf balls or polls covered with spines. Each grain contains one, sometimes two, male reproductive cells, which fertilize the female organs of plants, called stigma.

DIVERSITY

Not all types of pollen are useful for forensic purposes. Plants that self-pollinate do not need to disperse their pollen into the air or to manufacture much of it, so little is likely to be found on evidence from a crime scene. Plants whose pollen is spread by insects and other animals also are low producers and depend little on the wind. Their pollen, however, is long lasting and usually concentrated in specific areas, so it is a valuable forensic tool. Airborne pollen is abundant and widespread, which makes it easy to collect but also likely to be found in many places, unless it is a heavy type that sinks rapidly to the earth. In such cases, pollen rain, as it is called, does not widely disperse and can

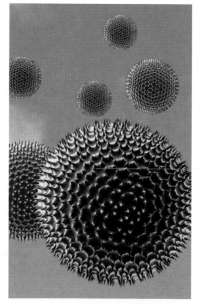

Left: Under the microscope, pollen resembles spiny-covered golf balls; except for pollen produced by self-pollinators, most types are appropriate for forensics evaluation. Right: Pollen is spread by insects or other animals, or becomes airborne to land upon and fertilize the female part of the plant, called the stigma. Top left: Anthers, the male part of a plant, is the site of pollen production.

quickly cover small areas, ideal for sampling and as an indicator. Fir trees produce heavy pollen, while juniper pollen is light.

COLLECTION AND EXTRACTION

Collecting the fine, powdery material that is pollen for forensics purposes takes exacting care. When the target is pollen within dried mud, for example, the sampler must remove any pollen that has later settled on the mud's surface—with a brush that has been scrupulously cleaned before use, for example. Dust-laden pollen can be sampled by pressing down on it with a strip of sticky tape. Experts advise the sampler to fold the sticky side of the tape back upon itself to protect the sample from contamination. Hair—human

or animal—acts like a magnet for pollen and can be a fertile source for forensic analysis. Pollen can be removed from hair by washing it—a mild detergent and distilled water is one recommended mix—and then freezing the wash water for storage until later examination.

Pollen is extracted from samples in the laboratory by a variety of methods, including dissolving the extraneous material, such as soil, with chemicals such as acetone, acids, and alcohol. Since the material being extracted is so small, palynologists use tiny containers, such as test tubes, some holding only a few milliliters. Optical microscopes with great magnification and scanning electron microscopes are among the instruments that palynologists use to analyze pollen.

Above: In addition to pollen, other plant matter—such as petals, fruit, leaves, or needles—can prove to be useful evidence. Top: The pollen of conifers comes from cones, which are more likely to spread via animals.

Soils Yield Secrets

What is solid rock today may someday be sand and another day in the future may again be rock. The Earth is never constant but is dynamic, always changing, although geological change may take eons. Wind, water, temperature changes, and human activity wear on rock, break it down, and carry the soils that result far and wide. Rocks, minerals, fossils, and the type and particulate sizes of soils vary immensely across the face of the planet and often quite distinctly, so that different locations have their own geological trademarks. More than 50,000 different soil types, according to the U.S. Department of Agriculture, lie underfoot in the United States alone. Forensic geologists try to characterize these differences in earth materials from place to place to identify locations that are significant to both criminal and civil investigations.

SOIL SAMPLES

Like so many other types of forensic evidence, samples of soils and other earth materials are valuable when they link a suspect to a victim and a crime scene. Geologists collect soil samples for forensics work much as they do for other purposes; however, they keep an eye out for factors that may have a significant forensic implication, such as soil moved by human activity, uncommon or unusual particles or minerals, and fossils. They use sieves and graduated cylinders to sort out soil particles according to size and to chemically separate particles cemented together by materials such as calcium carbonate and iron. Once samples are in the laboratory, soil particles and thin slices of rock can be examined using a petrographic microscope. Scanning electron microscopes help detect scratches and other features in minerals.

Right: Soil movement from human activity means presence at a crime scene. Above: Both petro-graphic and electron microscopes can be useful in the analysis of soil samples. Top left: The scientific classification of rocks is called petrography.

EVIDENCE IN THE GROUND

A wide spectrum of crimes have been solved with the help of secrets that forensic geologists uncover from soils and other earth materials. The type of rocks used as weapons have been traced to their source and have implicated suspects. Soil found encrusted under the fenders of vehicles and dropped on the road during impact in hit-and-run accidents has led investigators to the drivers involved. Pollutants dumped illegally have been traced to polluters. The locations of people wanted by the law or missing have been identified from rocks and other geological features showing in photographs in which they have appeared. And soils found on stolen agricultural crops have provided evidence leading to the arrest of thieves who stole them from the farms where they were grown. In one case, a man was found with a load of potatoes believed stolen from a particular farm whose soil had been heavily fertilized with phosphate, nitrogen, and potash. Nitrogen and potash quickly leach from the soil, but phosphate remains for a long period of time. When phosphate found in soil on the potatoes under investigation matched that in the soil of the farm, the case was solved and the potato thief convicted.

Top: Authorities can punish illegal dumping by tracing it back to the perpetrators. Middle: Forensic geologists may be able to match the soil on stolen produce with the earth of a particular field. Right: A rock used as a weapon can be traced to its place of origin.

Storm Stories

E arly on a December day, shortly before 3 o'clock in the morning, paramedics responded to a 911 call from a man in Durham, North Carolina, saying his wife was lying at the foot of the stairs in their home, bleeding from head wounds and unconscious. When the medical team arrived, they found that she was dead. The husband claimed that she must have fallen down the stairway.

After a coroner's inquest and investigation, prosecutors charged the man with murder. Two years later, after a much-publicized trial featuring hours of testimony on forensic evidence offered by both prosecution and defense, the husband was convicted of bludgeoning his wife to death. Among the evidence gathered by the prosecution in the case was that from a forensics meteorologist regarding weather conditions at the outdoor pool where the husband claimed he had been between 11 P.M. and 2 A.M., before he made the call. The meteorologist went back into meteorological records and reconstructed a picture of what he thought the weather was like when the accused husband claimed he had been outside by the pool. A cold front had

Above: A man who claimed to be lounging by the pool in a T-shirt and shorts was later convicted of murdering his wife when a forensic meteorologist testified that the weather was too cold. Right: Forensic meteorologists look back in time to determine if weather conditions match a suspect's testimony. Top left: The phase of the moon and cloud cover can determine light on a particular night and how well a witness could view an event.

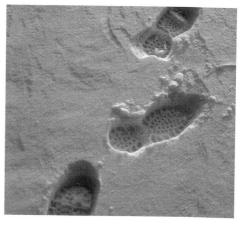

moved through and the temperature had ranged between 51 degrees and 55 degrees, according to the assessment, chilly for sitting at poolside, especially in shorts and a T-shirt, which the husband had been wearing.

METEOROLOGICAL MATTERS

Forensic meteorologists are called in many types of criminal and civil cases to reconstruct weather conditions at the time the crime, accident, or other event occurred. They

Left: Analyzing atmospheric conditions such as fog and wind can help determine how far sound can be carried. Above: The National Climatological Data Center can provide such helpful items as radar images and satellite photographs.

may be asked, for example, to determine whether snow had melted sufficiently to alter a footprint allegedly left by a suspect, whether the sun was at a position in the sky to cause glare that might have contributed to an automobile accident, whether atmospheric conditions such as fog and wind affected the distance at which cries for help could be heard. The ambient temperature when and before a body was found could help pathologists determine how long the individual had been dead. The phase of the moon and prevailing weather conditions on the night of an event may determine the accuracy of testimony by someone who allegedly witnessed it.

READINGS FROM THE PAST

Unlike the meteorologist whose job is to predict weather, the forensic meteorologist looks back in time. Nevertheless, the standard tools of the weather trade are used. Satellite data, forecasts, advisories, and radar images may all be involved in a case and, increasingly, forensic meteorologists call on computer graphics to develop and present their analyses. They also examine a region's climatic patterns and weather trends.

One problem for these weather sleuths is that meteorological observations were often taken at sites far removed from the location at which the investigation is taking place. Often, reports from several weather stations in a region must be compared and their similarities and disparities are factored into the analysis. Frequently, as well, a forensic meteorologist must analyze the weather impact on very small areas, called microclimates. Immediate factors, such as heat from nearby buildings, topographical elevation, and groundwater levels, may have to be accounted for, and, if the event under investigation took place on an oceanfront beach, tides may be involved.

The phases of the moon will affect tide cycles, which might prove helpful in determining the movement and placement of evidence at a waterfront crime scene.

Insects Tell Time

Flies, with their whiny buzzing and penchant for landing on the potato salad set out for an afternoon picnic, annoy most people, but not forensic entomologists. They prize flies, or at least certain species. The behavior and, especially, the life cycles of a small number of flies and some other types of insects that feed on decomposing human tissue can be correlated with often astonishing accuracy to when a person died, whether a corpse has been moved, and sometimes even how that individual died.

BLOWFLY SPECIES

The most important of all insects to forensic entomology is the blowfly, more properly the blowflies, because there are several species. Blowflies, sometimes called greenbottle or bluebottle flies because of their coloration, lay their eggs in dead flesh. Maggots are the larval stage of these flies that hatch from the eggs and feed upon decomposing tissue.

Blowflies are near-perfect indicators for forensic entomologists. They are ordinarily the first carrion-eating insects to show up at a dead body, usually within a day or two of death. Their life cycle, including the rate at which the larvae grow, pupate, and metamorphose

Above: Blowflies include the bluebottle, the screwworm, and the greenbottle, pictured here. Top left: The life cycle of the blowfly provides a very precise time sequence for analyzing dead bodies.

into mature flies, has been precisely calculated. The speed at which growth occurs depends to a degree on ambient temperature, cloud cover, wind, and other weather conditions and even the type of earth materials of the surroundings, so forensic meteorologists and geologists sometimes lend entomologists a hand with their calculations. Entomologists also often work alongside forensic anthropologists and pathologists to gather

Above: Various factors considered when determining blowfly cycles include ambient temperature, cloud cover, and wind. Below: This larval stage of the blowfly evolves in about a day from an egg laid in an abundant food source, which can include a dead body.

the facts about a corpse from the perspectives of different scientific disciplines, thus creating a multifaceted picture of the circumstances involved in a homicide or other crime.

BREEDING BEHAVIOR

Homing in on the foul odor of a decaying body, the female blowfly lays her eggs on it, usually near or in openings such as the nostrils, eyes, ears, and anus, or, when the situation applies, wounds. The eggs hatch quickly and the larvae start to feed, in a manner that probably should not be discussed at the dinner table: Their hooklike mouthparts chew away at tissue that is softened not just by decomposition but by enzymes in the insects' excrement. The larvae grow through three stages, called instars, shedding their outer skin between stages, then transform into pupae; this process takes about two weeks.

STAGES OF LIFE

The mouthparts and some other features of the larva that change with growth, as well as the larva's length, are indicators of its age. Typically, the larva starts life at 0.08 inch (2 mm) long. In slightly less than 2 days it is 0.2 inch (5 mm) long, and reaches 0.4 inch (10 mm) long at 2.5 days. It then becomes an eating machine and its growth is explosive. By 5 days it has a length of 0.7 inch (17 mm). At 8 to 12 days it starts the transformation into a pupa, stops eating, starts moving away from its food source, and begins to shrink, to

The cocooning of the fly larvae into an adult fly typically takes from 3 to 5 days, providing a reliable time frame for forensic scientists. From egg to death, the entire life cycle of a blowfly is between 9 and 10 weeks.

about 0.5 inch (12 mm) long. Covered by a hard cocoon shell, the pupa shrinks further, to 0.35 inch (9 mm). Between 18 days and two weeks, the change to adulthood occurs and the newly metamorphosed fly wings away, leaving its empty shell behind. These facts, and even the empty shells provide entomologists with an estimate of how long a body has been in a particular location, a distinct advantage over relying on indications of rigor mortis and algor mortis (the gradual cooling of a body after death), which disappear after about 5 days.

Insects in Investigations

It is a grisly sight: blowfly maggots feeding on the flesh of a decaying corpse. Their soft bodies moving grotesquely and ever so slowly, they seem everywhere. They may seem everywhere, but the practiced eye of a forensic entomologist can quickly spot the areas in which maggots have concentrated. Some of these areas can help investigators deduce the way in which the deceased individual met death.

EGG DEPOSITS

Blowflies looking for a site on which to deposit their eggs favor the face, because it has the largest number of natural openings: nostrils, eyes, mouth, and, nearby, ears. They also head for the genital area and anus. If, in the estimation of a forensic entomologist, there are a greater number of larvae in the region of the anus and genitals than is usual, crime of a sexual nature may have preceded death. Bleeding from these areas makes them even more attractive to the female flies.

Abnormal concentrations of blowfly maggots also can suggest the manner in which a victim of violent homicide died or the events surrounding the attack, because wound openings also attract egg-laden females. A mass of maggots at the rear of the skull may indicate that the victim suffered a blow from behind. If large numbers of maggots are feeding on parts of the hands and arms, the victim may have been cut in these places while trying to ward off a knife or other weapon wielded by his or her attacker.

CLUES FROM MAGGOTS

Maggots may also prove useful in an investigation that is hampered by severe decomposition and other circumstances that hinder detection of drugs and poison in body tissues and fluids. When maggots feed, they

Above: Maggots are attracted to blood and are reliable indicators of wounds. Female blowflies will seek out wounds as sites to lay their eggs. Top left: Site-specific blowflies help determine if a body has been moved from the scene where the homicide took place.

Toxins in the body, such as cocaine and heroin, will slow down the timing of larval development, prolonging the process and leaving trace elements of the drugs in the post-larval cocooning shell.

chemicals, including cocaine, extend the time in which larvae develop. Creatures avoid certain chemicals, which would cause them to stay away from orifices where they normally gather.

Insects can even suggest whether a body has been transported after death from one place to another. Some species of blowflies like sunny habitats; others prefer shady places. Some are country blowflies, living in suburban and rural areas. Others cope quite well with urban conditions. If blowfly maggots intolerant of an urban habitat are found on a body dumped in a city alley, the victim probably was killed out of town and then transported to the place where it was discovered.

OTHER INSECTS

Along with blowflies, a succession of other insects may visit a dead body, some to feed on the corpse, others on the insects that are eating it. Flesh flies arrive on the heels of blowflies, but they leave behind their larvae rather than eggs. Later, predatory flies arrive to feed on maggots, along with other insects such as carrion beetles and mites, not insects but, like them, arthropods. Comparison of insects found underneath the corpse with those already on it can provide an indication of whether the body has been moved.

may ingest foreign chemicals along with the decaying tissue. These chemicals may be extracted from the maggots; one method is to extract chemicals from a puree made of maggots. Traces of drugs and poisons may also be found in the shells cast off by larvae as they grow. Researchers have found that certain chemicals can influence the development of maggots and even where they congregate on the body. Some

CHAPTER 11

DIGGING INTO MYSTERIES

Left: Forensic experts gather the skulls, bones and clothing of victims unearthed in what may be the largest mass grave ever found in Bosnia. Top: Two beams of light were lit at Ground Zero in lower Manhattan on the third anniversary of the 9/11 terrorist attacks. Bottom: In 2004, a 9.0 earthquake in the Indian Ocean triggered a tsunami that spread over more than 10 countries in Southeast Asia and northeastern Africa.

Death investigations in which forensics are involved usually center on one, a few, or at most a handful of people. Yet increasingly, and unfortunately, disasters both natural and man-made result in mass fatalities that demand forensic investigation. The causes of mass fatalities are myriad: terrorist attacks such as the destruction of the World Trade Center; ethnic cleansing and other variations of genocide as in the former Yugoslavia and Rwanda; airline disasters, whose death tolls increase as larger airliners are engineered; and earthquakes, tsunamis, super hurricanes, and other natural disasters that can kill hundreds of thousands in one event. Many of the forensic techniques used when massive numbers of deaths occur are the same as those used when the death toll is only one or two people. Other techniques are also required, adapted to coping with the sudden crush of bodies and conditions that make it almost impossible to work effectively. Communication between the investigators who perform traditional forensic investigations and those who use newer ones tailored to mass disasters means that improvements in either methodology benefit both.

Who Killed King Tut?

FORENSICS AND ARCHAEOLOGY

Whether they are unearthing clues to mysteries of the present or of secrets from the past, forensic anthropologists usually examine artifacts such as tools, weapons, personal effects, and other material remains along with the remains of the people who used them. When digging up the fossils or burial sites of prehistoric peoples, for example, forensic anthropologists are in effect uncovering stones as well as bones. Often these scientists have practical experience dealing with material remains but, increasingly, scientists who are specifically trained to deal with them have entered the field of forensics.

FORENSICS AND ARCHAEOLOGY

Because the tasks of excavation and recovery are the archaeologist's stock-in-trade, these scientists are frequently called upon to locate remains and help uncover them. In fact, a standard archaeological dig depends upon techniques virtually identical to those used in forensic investigation of remains hidden in graves and similar locations. Forensic archaeologists are particularly important in investigations of mass graves and other scenes of war crimes, as well as ancient mysteries such as the never-ending debate over who—or what—killed King Tut.

WAS IT MURDER?

Examinations of the world's most famous mummy by forensic experts and Egyptologists have resulted in a variety of opinions about how and why the teenaged Egyptian pharaoh died, but most suggest he was murdered. Over many decades these opinions have waxed and waned in favor and have prompted a host of television programs, books, and articles on the subject.

Tut—Tutankhamen—gained notoriety not because of his achievements while on the throne but because his tomb, uncovered in 1922 by Englishman Howard Carter, was the most intact and replete with riches ever

Above: Tablet at the entrance to the tomb of King Tut, which was first uncovered in 1922. Top left: Whether investigating the mysteries of ancient peoples or solving a more recent crime, archaeologists rely on similar tools and methodologies.

The 1920s damage to Tut's mummy is difficult to distinguish from damage dating back to the king's lifetime.

discovered. Unfortunately, Carter and his team dismembered much of the mummy in their efforts to retrieve artifacts and remove the body from its sarcophagus. Tut had been affixed to his coffin by the resins and other fluids used in the embalming process. Nonetheless, his 3,300-year-old remains have been subjected to innumerable forensic tests, including virtual bone-by-bone examination as well as X-rays, as have paintings and artifacts found along with it. Damage to Tut's skull has been the key piece of evidence used by proponents of the idea that he was murdered. An X-ray of his skull revealed a calcified blood clot at its base. This could have been caused by a blow from a blunt implement, which eventually resulted in death. Other investigators of this death on the Nile theorize that the king was poisoned and that the skull injuries shown in X-rays may have happened when his body was dropped. Whatever the exact cause of death, murder-theory proponents speculate that his elaborate tomb may have been a cover-up. Wall paintings seem to have been executed sloppily, even at the last minute, and funerary items may have come from a secondhand store. Names of other people seem to have been erased from them.

NEW RESEARCH INTO AN OLD MYSTERY

Early in 2005, an Egyptian-led research group came up with another solution to the mystery. Computerized CAT scans of the mummy, they said, show that the thigh bone of his left leg had been broken before death and, the researchers claimed, showed no evidence of a blow to the head. Their theory of what killed the teen king: lethal infection stemming from a severely broken leg. As with all other theories about the death, it was challenged. If there is one firm conclusion that may be drawn from the myriad probes of who—or what—killed King Tut, it may be that we will never know.

Above left: Tomb entrance in the Valley of the Kings, Luxor, Egypt. Above right: Scholars have speculated that because many tomb furnishings are of inferior quality or secondhand, it may indicate a hasty burial for the murdered king. Below: The CT scan done on King Tut's mummy indicates that he was not murdered after all, but may have suffered a badly broken leg shortly before his death.

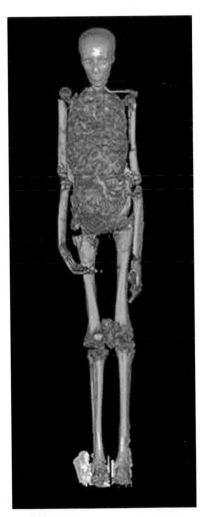

The Iceman Mystery

Although lacking the glittery, golden aura of King Tut, and mummified by nature rather than by ancient embalmers, the ancient Iceman of the Italian Alps has also been the subject of voluminous media attention. Actors in one television program even re-created a battle that may have been responsible for the defensive cuts on the Iceman's hands as he warded off the sharp-edged weapons of opponents. From all indications, he was a fierce and skilled fighter, who probably took a few of his opponents with him.

WHO WAS THE ICEMAN?

Unlike Tut, who sat on the pharaonic throne of ancient Egypt, the Iceman, discovered high in a pass on the Schnalstal glacier of the Tyrolean Alps in 1991, was a common man, perhaps even an outcast. After several years of forensics testing on the 5,300-year-old mummy, nicknamed Ötzi from the Öztal Alps where he was found in a very well-preserved state melting out of the ice, scientists believe that they now know how he died. Apparently, Ötzi was attacked and managed to flee. As he ran he was shot in the back with an arrow. Although he or someone else managed to pull out the arrow shaft, the head remained stuck in his shoulder. When he reached the top of the mountains, he could go no farther. Exhausted and weakened from loss of blood, he lay down and died. The cuts on his hands and an arrowhead in his back support the fight theory, but what led to the battle and why Ötzi fled into the mountains still eludes and tantalizes researchers.

AN AMAZING DISCOVERY

Ötzi was found, facedown in the ice and intact, by two hikers just a few strides from the Austrian border. He probably was in his thirties or forties, was about five feet three or four inches in height, and wore a woven grass cloak, leather vest, waterproof shoes of bear and deer hide with grass "socks," a belt from which to drape his loincloth and suspend his leggings, a jacket, a cape, and a bearskin hat. He carried a wooden-framed

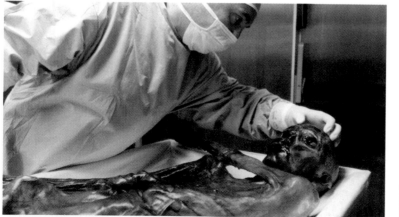

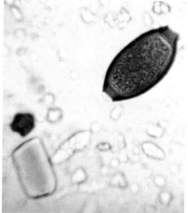

Left: A researcher prepares to take DNA samples from the body of a Bronze Age hunter known as "Ötzi" for the alpine location in which he was found. Right: An egg from the "human whipworm," Trichuris trichuris. Ötzi's stomach showed signs of parasitic whipworms. Top left: The glacial conditions of the Italian and Austrian Alps preserved the Iceman for 5,300 years after his death.

backpack, two bark containers, one containing charcoal, and a belt pouch housing small useful items including flints, a retouching tool, and fungus for tinder. He was well armed, with a hafted copper ax, a flint knife, arrows in a quiver, and a bow as big as the English longbow of the Middle Ages. He had 57 tattoos, located at what seem to be the points used in acupuncture.

The Iceman was put through the full gamut of forensic testing. The contents of his intestinal tract consisted of mostly wheat, probably ingested in the form of bread; researchers could therefore deduce that he lived in an agricultural culture. He consumed his last meal about eight hours before he died. Electron microscopy of pollen in his stomach indicated that he died in the spring. He had parasitic whipworms, most likely from food or water contaminated by human waste. Chemicals in his tooth enamel suggested that he spent most of his life in valleys to the south of where he had been found.

A CAT scan revealed the arrowhead in his back; its location indicated that someone else had broken off the shaft. A tear in his cloak matched the site of the arrowhead. Blood on his weapons and coat were from four other people, probably opponents or companions in battle. Blood on one of his arrowheads hinted that it had hit two different foes. All in all, forensics has portrayed the Iceman much as he was, a citizen of Copper Age Europe.

Although she was frozen in the frigid temperatures of Mount Ampato in Peru, the mummified body of "Juanita," also known as the "Ice Maiden," was only discovered because a nearby volcano had caused Ampato's snowcap to melt.

ICE MUMMIES

Ice mummies, or bodies preserved by cold, have been found in several parts of the world, notably the Andes of South America. Many Andean mummies are those of young women and children, dating to Inca times and even much earlier. Perhaps the most famous of these is known as "Juanita," found high on Mount Ampato, Peru, in 1995. No more than 14 years old, she was apparently sacrificed about 500 years ago. Near her body were statues and artifacts, believed to be offerings to the gods. Examination of the mummy showed that she had been killed by a blow to the head. Juanita's body was exposed by chance. Hot ash from the eruption of a nearby volcano melted the ice that had concealed her. She was discovered before winter snows arrived to hide her once again.

CAUSE OF DEATH

What has not been determined is the reason for the Iceman's death. Was he fleeing a conflict between rival clans over territory or a raid for slaves and booty? Was he driven from his village for some offense? Early on, there was speculation that he was a ritual sacrifice, perhaps to a fertility deity, but that idea was generally discredited. On the other hand, fertility may have had something to do with the Iceman's death. Early in 2006, results of tests on the Iceman's mitochondrial DNA (DNA found outside the cell nucleus and inherited maternally) were released. Mutations were found in the DNA that are often linked to male infertility. Childless, without a family, Ötzi the Iceman may have been an unwanted loner, on his own when he most needed help.

Cannibal Cousins?

Like modern humans, the Neanderthals who lived in Europe and western Asia until about 30,000 years ago were capable of both tenderness and savagery toward their fellows. Anthropological and archaeological evidence from caves where Neanderthals took shelter from the Ice Age elements show that they sometimes buried their dead with flowers and cared for their infirm. Analysis of pollen at a famed grave site, in Iraq's Shanidar Cave, revealed that groundsel, thistle, mallow, and several other flowers had been placed there. One skeleton of a Neanderthal male who died at the ancient age of 45 showed that he had a withered and handless right arm and a badly injured collarbone and legs. To survive, especially to such an advanced age for the time in which he lived, his fellows must have nurtured him.

A VIOLENT SOCIETY

The Neanderthals, however, who branched off from the modern human lineage about 500,000 years ago, were far from flower children. Analysis of skeletal remains indicates that, as do humans today, they sometimes fought violently. Several skeletons show stab wounds on the left side of the body. Excavation of Neanderthal caves has also revealed the scattered bones of numerous individuals jumbled up with the bones of animals that served as food. Among them was a huge Neanderthal site in Croatia, Krapina Cave, first excavated in the late nineteenth and early twentieth centuries. Such findings sparked debate over whether Neanderthals ate other Neanderthals.

Above: Model of a group of Neanderthals during a burial ceremony. Although they buried their dead and looked after the infirm and the disabled, evidence from bones found in caves suggests that Neanderthals may have also practiced cannibalism.
Top left: Flowers such as thistle have been found buried with Neanderthal corpses.

Skeleton of a Neanderthal man taken before entering a famous cave, La grotte de Clamouse, in Languedoc in the south of France. Neanderthals were early humans that lived in Europe and the Middle East about 120,000 to 30,000 years ago.

other food was scarce or as part of a social ritual. Yet, the way in which the bones were butchered and the fact that animal bones were handled in similar fashion suggest that some of our ancient cousins were, in fact, cannibals, a conclusion that may seem at odds with the evidence that Neanderthals carefully buried their dead. According to Professor Tim White of Berkeley, such variable treatment reveals a cultural complexity, "a pattern that mirrors the behavior of more modern people."

A reconstructed Neanderthal skeleton, right, stands alongside the skeleton of its distant cousin, the modern human.

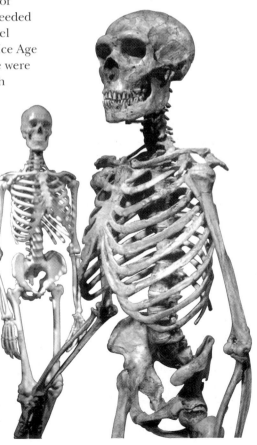

SUGGESTIVE EVIDENCE

The question was answered, at least in part, in 1999, when a team led by Alban Defleur of the Laboratoire d'Anthropologie in Marseille, France, announced findings from a dig at the Moula-Guercy cave, located near that city. Defleur and his team retrieved bones from at least six smashed Neanderthal skeletons along with animal bones, in a recovery effort that a colleague called "treating the site like a crime scene." More than 70 bone fragments, some just tiny bits and slivers, were examined. Several turned out to be parts of a Neanderthal femur with cut marks from stone tools used to deflesh it, and also marks from a stone hammer and anvil that had smashed it open. Bones of the red deer were treated in the same way. No signs of gnawing were found on the bones, ruling out the possibility that wild animals ate the Neanderthals. The verdict: The

bones were smashed to expose the marrow, a rich source of protein, which was badly needed by the Neanderthals to fuel their bodies under harsh Ice Age weather conditions. There were no signs of charring, which suggests that the flesh was either eaten raw or cooked off the bone.

Despite the evidence, researchers cannot say for sure that cannibalism was the rule for Neanderthals, or that it was always a means of sustenance. The people of some cultures have eaten the remains of killed enemies as a means of gaining their power or consumed deceased relatives as a form of communion with their spirits. It is unclear from existing evidence whether individuals were eaten for survival when

The Bog Man Murders

Throughout northern Europe, from Ireland to Scandinavia, hundreds of bodies dating to Europe's Iron Age, around 2,000 years ago, have been found preserved not by freezing temperatures or by the embalmer's art but by the acids of peat bogs that are common in that part of the world. Many of the people whose bodies have been found, mostly men, were tortured in grisly fashion before they died. A number of these preserved corpses, or "bog mummies," seemed to be of victims of ritualistic murder who were killed in similar ways, creating a pattern of sorts. This is not to say that they were the victims of an Iron Age serial killer who roamed northern Europe in search of victims. It appears that in many cases, the victims were human sacrifices to harvest and fertility gods. Some of the sacrifices may have served a dual purpose: to propitiate the gods and execute criminals.

WHO WERE THE BOG MEN?

The bog man murders have been investigated using many of the forensic principles used in modern homicide investigations. In fact, the discovery of a man who died in an Irish bog prompted police to search for a contemporary killer until archaeologists looked it over and reported that the body, exceedingly well preserved, was more than 2,000 years old. This particular body, known as Old Croghan Man after the name of the place it was found, had fingerprint whorls as visible two millennia after his death as they were during his lifetime.

Many of the analytical techniques that have proved impossible to use on other ancient remains worked in the examination of the bog mummies. The combination of acids that saturate the soil of peat bogs is a remarkable natural preservative, keeping intact skin, intestines, brains, hair, and other soft tissues that otherwise would decompose and disappear. The skin of bog mummies has been blackened and is leathery, and the bodies also take on a stretched-out appearance.

Bog mummies have undergone X-rays, autopsies, CT scans, bone examinations, analysis of stomach contents, and many other forensic procedures in efforts to find out about how they lived and how they died. The last meals of many bog men were composed of what must have been noxious-tasting soups made from plants. These may have been ritual meals because, in at least some cases, they do not appear to have been part of the regular diet.

Above: Of the hundreds of bodies discovered, some have been remarkably preserved by the acid- and oxygen-free conditions of the bog. In many cases, skin and clothing have remained intact over the course of 2,000 years. Top left: Field of barley, an ingredient in the soup that was the last meal of many of the bog mummies.

CAUSE OF DEATH

Most bog men (and women) died from strangulation, slit throats, or drowning, sometimes in combination. Leather ropes have been found around the necks of some of these peat-preserved bodies. Old Croghan Man and another Irish bog man, both described in 2006, suffered horrendous deaths. Old Croghan Man was stabbed, his nipples were sliced off, rope was threaded through his arms, and he was beheaded. Clonycavan Man's head was axed three times and he was disemboweled.

UPPER-CLASS VICTIMS

The two Irish bog men, like many others, show signs that they belonged to the upper class of their society. Their hands are smooth, not roughened by manual labor. Old Croghan Man's fingernails were manicured. Clonycavan Man's hair was spiked with a resinous gel made from resin from pine trees found in France and Spain, a costly imported product. The hair and beard of Lindow Man, a bog man found in England,

revealed interesting details, including the fact that he sported a beard. No other male bog body had been found with a beard, which were rare during the time he lived. It was also clear that someone had trimmed Lindow Man's hair with scissors two or three days before his death. Historians and archaeologists knew that, although scissors existed in England at the time, they were reserved for a privileged few. Yet, Lindow Man was found naked, and without clothes to provide further clues, researchers could only speculate about whether he was an aristocrat or dignitary of some sort. The fact that the upper classes provided some of these apparent sacrificial victims may mean that wealth and power were no shields against the perceived needs of the gods. Alternatively, wealth and power might have been thought to enhance the offering. Many remains are buried

Bog men used hair "product" made of Spanish and French pine resin.

on important political or royal boundaries. Ned Kelly, keeper of Irish antiquities at the National Museum of Ireland, postulates why this may be: "My belief is that these burials are offerings to the gods of fertility by kings to ensure a successful reign. Bodies are placed in the borders immediately surrounding royal land or on tribal boundaries to ensure a good yield of corn and milk throughout the reign of the king."

A bog mummy named "Red Franz," a man in his late twenties when his throat was slashed almost 2,000 years ago and buried in a bog in Germany.

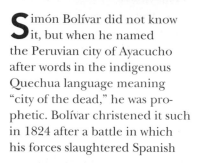

Mass Graves and Human Rights

Simón Bolívar did not know it, but when he named the Peruvian city of Ayacucho after words in the indigenous Quechua language meaning "city of the dead," he was prophetic. Bolívar christened it such in 1824 after a battle in which his forces slaughtered Spanish troops. The name took on a grimly ironic significance in the 1980s and 1990s. Nine thousand feet in the Andes, on terraced slopes below the brooding, barren landscape of the Peruvian altiplano, Ayacucho was the focal point of atrocities by Marxist Shining Path terrorists and responding torture and terror by government-backed death squads that killed more than 69,000 people.

A SPECTER OF VIOLENCE

The government response crushed the Shining Path but led to the ouster of hard-line Peruvian president Alberto Fujimori in 2000. Even so, the specter of violence still hovered over the city. Early in 2005, three Peruvian forensic anthropologists exhuming graves suspected of holding Shining Path atrocity victims received text messages threatening their lives. So serious was the threat that Dr. Elizabeth Brumfiel, president of the American Anthropological Association, wrote the government of Peru demanding protection for the forensic scientists. It was not the first time that she had found it necessary to demand protection.

Above left: Forensic anthropologists from Peru's Truth Commission diligently catalogue remains of human rib bones taken from a mass grave site in Totos, southeast of Lima, Peru. Above right: Forensic anthropologists often work with organizations such as the United Nations to uncover atrocities. Top left: Symbol of the Shining Path, a ruthless terrorist group founded in Peru.

Crania found in the mass graves in the Killing Fields of Phnom Penh, Cambodia.

play a role in identifying victims of atrocities, although the deterioration of and damage to remains can thwart recovery of sufficient genetic material. The challenges facing DNA experts in places such as Bosnia have led to improvements in DNA testing technology. American scientist Edwin Huffine developed a computer-linked identification program that extended the source of DNA samples for matching to even distant relatives of the deceased. A new method of extracting DNA from teeth was developed for 10-year-old remains from the civil war in Guatemala. A huge DNA database set for victims and their relatives in Iraq identified thousands of remains and also proved that earlier identifications were wrong by a factor of half.

The organization has often complained about threats to anthropologists working on mass graves and other dangerously sensitive human rights issues around the globe.

IN THE THICK OF THINGS

Forensics anthropologists and, indeed, forensics in general are in the thick of the effort to document human rights violations by detecting mass graves and other atrocities and examining the remains found therein. They have operated in Iraq, Argentina, Rwanda, Guatemala, Bosnia, Kosovo, Columbia, and many other countries to unearth evidence of disappearances, mass executions, and torture, often in conjunction with the United Nations. When examining evidence of mass killings,

forensics experts pursue goals similar to those in a homicide investigation. The need to show cause of death, time since death, and other circumstances surrounding the crime is the same whether the victim is a single individual or the dead number in the hundreds, even thousands. When many bodies are in the same grave, however, there are additional questions, such as how many individuals are represented. Sorting out the number requires separating victims by sex and age, which can be extremely difficult, especially when victims have been exhumed and reburied to conceal their location.

DNA LEADS THE WAY

Pathology, examination of bones and teeth, and DNA profiling all

Deterioration of genetic remains can make identification nearly impossible, but advances in DNA technology are enabling positive matches with even severely damaged samples.

The World Trade Center

In February 2005, the New York City medical examiner's office effectively hung up a "Closed" sign on the door of its efforts to identify the remains of people killed when two planes crashed into the World Trade Center on September 11, 2001. Although it was no solace to the families of the victims who were still unaccounted for, the remains of more than 1,500 people, about half of those lost, had been identified. Of the 20,000 bits of tissue and bone recovered, most small enough to fit into a standard laboratory test tube, 10,000 remain unmatched.

Above: A cloud of smoke and dust rises from one of the towers of the World Trade Center. Efforts to identify the remains of those trapped inside began immediately and continued for years afterward. Top left: A makeshift memorial to the lost in lower Manhattan.

AN OVERWHELMING UNDERTAKING

The success rate of the identifiers was nevertheless remarkable given the crushing magnitude of the task and the conditions under which scientists worked, especially during the earliest days after the disaster. A forensic dentist from the University of Texas Health Science Center described typical work at the site. He spent 12-hour shifts in a makeshift tented morgue near the blast scene. All day long, bulldozers and excavators snorted and rumbled as they dug through mountains of debris a few hundreds yards away, where fires still burned and smoke clogged the air. Progress was painfully slow; on one particularly grueling shift, only eight bodies were processed. A team of volunteer anthropologists from Brown University in Providence, Rhode Island, and Brooklyn College in New York City, worked for two days, sifted through 90 buckets of soil, ash, and other debris, and found only 10 small pieces of human bone.

Many of the victims were virtually vaporized by the initial blasts, eliminating any chance of identification. Even many of the bits and pieces that were found were compromised as far as DNA testing is concerned by the

1,800°F (982°C) heat of the fires that enveloped both towers of the trade center. To further complicate issues of identification, unlike airline or ship disasters, in which there are generally passenger lists that provide finite lists of possible victims, there was only an estimate of how many people could have been in the towers at the time the terrorists hit. Added to the 50,000 people who worked in the towers, an estimated 90,000 visitors headed there every day—tourists who added New York's tallest building on their lists of must-see attractions and shoppers at its indoor mall and businesspeople from around the globe. At 8:46 A.M., just who had been in the North Tower when the first plane hit? Who was in the South Tower just over 15 minutes later when the second plane crashed into it? There was no easy way to check off who was missing and who was accounted for.

NEW WAYS TO ACCOUNT FOR THE MISSING

The trade center disaster sparked new thinking about how to handle forensics recovery when mass deaths occur. Several panels were established to evaluate methods and techniques. One, convened by the National Institute of Justice, spurred the development of improved computer software that works faster and better integrates information needed for DNA profiling.

Since September 11, emphasis has been placed on newer DNA techniques, including a variation on the standard

short tandem repeat tests than can provide results for matching from shorter, more degraded pieces of DNA. Another effort focuses on using only a single location on a piece of DNA for telltale genetic variations. New DNA databases, such as the Mass Fatality Identification System, have been developed. Along with new testing methods, matching would be far more effective if DNA of lesser integrity could be used. An option under development is synthesizing enzymes that naturally repair DNA and using them to reconstruct damaged samples.

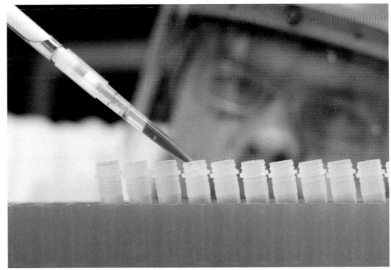

Above: A forensics researcher analyzes samples. In the aftermath of the 9/11 attacks, new technologies were developed for evaluating DNA in order to increase identification rates in the event of catastrophes involving mass deaths. Top: Flowers and other tokens of remembrance were left at Ground Zero for months following the terrorist attacks.

The Old-Fashioned Way

Sometimes the old-fashioned way works best. In the aftermath of the December 26, 2004, destructive tsunami that killed 155,000 people across the Indo-Pacific basin, experts scrambling to identify the bodies left in the wake of the catastrophe more likely than not were forced to rely on time-tested forensic methods of identification rather that new techniques such as DNA matching.

OBSTACLES TO IDENTIFICATION

Identification of remains by DNA was used to varying degrees of success after the World Trade Center disaster and in documenting mass graves in the former Yugoslavia, but in Asian countries devastated by the giant waves, finding appropriate laboratory resources and locating relatives to provide DNA samples for matching often proved nearly impossible. The hot, humid weather, moreover, increased the rate at which the bodies of victims decomposed, breaking down nuclear DNA quickly. As things turned out, the identification of victims depended largely on techniques used long before the advent of DNA testing, such as forensic dentistry, which proved extremely reliable under the trying conditions after the tsunami.

Top: Photographs of tsunami victims were posted in public places in the hope that friends and family would recognize and identify their loved ones. Bottom: Entire villages were destroyed as a result of the tsunami, as was this home in Phuket, Thailand. Top left: Typhoons can produce large, destructive, life-threatening waves.

TRIED-AND-TRUE METHODS

According to the respected magazine *New Scientist*, dental records identified 75 percent of bodies recovered in Thailand, fingerprinting 10 percent, and DNA profiling less than 1 percent. "Dental Records Beat DNA in Tsunami IDs," read a headline over the magazine's report. Forensic dentistry teams from New Zealand played a major role in the identification operation. Almost immediately after the disaster, dentists from around the world began sending dental records of patients who were reported missing to an international database. This process worked most successfully on missing people from Western nations than on those in impacted areas because of low levels of dental visits in many of these places. In some cases, members of dental teams identified victims by comparing their teeth with photos in which they were smiling.

OVERCOMING THE TSUNAMI'S CHALLENGES

The difficulties of identifying remains under the horrendous conditions at tsunami sites dramatized the advantages and disadvantages of different forensic identification techniques. Some laboratories that otherwise would have been capable of DNA work were either destroyed or incapacitated by factors such as road destruction and power outages. With no electricity, refrigeration needed to store DNA samples was at a premium, if available at all. Teeth, on the other hand, required no refrig-

Top: Volunteers placing bodies into coffins in preparation for cremation after the 2004 tsunami in Indonesia. Bottom: The terrible damage caused by natural disasters makes identification of remains even more difficult than in other circumstances.

eration for storage and were not subject to rapid decomposition. Bodies, however, decayed rapidly under the sun and humidity, bloating within a day or two, skin peeling off and discoloring. A forensic dentist from New Zealand's University of Otago said of bodies on a beach in Phuket, Thailand, that it was "difficult to tell whether they are Caucasian or Asian," much less determine their individual identities. In many places, investigators had to sort out individuals one by one from tangles of

corpses. Even bodies that were in relatively good condition posed identification problems, because many victims, especially foreign tourists, were on beaches in bathing attire without any sort of personal identification on them.

In many cases, bodies that were examined by pathologists and subjected to DNA testing still had to be given to a forensic dentist for final identification. Dentists were often the final arbiters who linked names with tsunami victims before they were repatriated for burial.

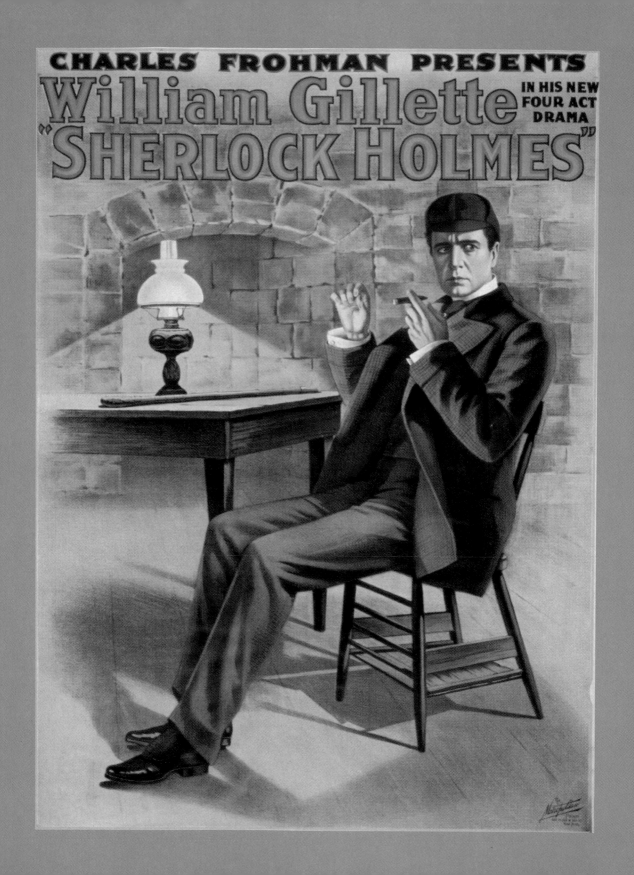

CHAPTER 12

FORENSICS, FACT AND FICTION

Left: The popularity of Sir Arthur Conan Doyle's fictional detective Sherlock Holmes, introduced to the reading public in 1887, presaged the contemporary appetite for forensics-themed books, movies, and television programs. Top: The red jellyfish, or lion's mane, features significantly in Conan Doyle's "The Mystery of the Lion's Mane." Conan Doyle hobnobbed with many scientists engaged in the emerging field of forensics. Bottom: A rocky beach on the Sussex coast of the English seaside, where Conan Doyle set his short story.

" '*Cyanea!*' cries Sherlock Holmes, while looking down at what appears to be a glob of pinkish glue lumped on the bottom of a tide pool. '*Cyanea!* Behold the Lion's Mane.'" The world's most well-known fictional detective used his knowledge of science to solve the mysterious death of a science teacher who perished after swimming on the seashore near Holmes' Sussex coast villa. A suspect emerges: a mathematics teacher who had been a romantic rival of the deceased. In the end, Holmes exonerates the suspect by revealing the highly toxic red jellyfish, or lion's mane—known in scientific terms as *Cyanea capillata*—as the culprit.

"The Mystery of the Lion's Mane" is just one Sherlock Holmes story in which the detective uses science as a tool of criminal investigation. Sir Arthur Conan Doyle was a serious student of the emerging field of forensic science. Conan Doyle, in fact, promoted forensics' use in criminal investigation by popularizing it almost ahead of its time. Today the media have again seized upon forensics to grasp the public's imagination, with implications for the legal system for better or worse. Popularization of forensics has both fostered acceptance of scientific techniques in police work and skewed public perceptions of the legal system.

Forensics Foreshadowed

The timing could not have been better for all concerned. Forensic scientists such as Sir Francis Galton, who pioneered fingerprinting as a means of identifying criminals, were trumpeting the use of science as an indispensable investigatory tool. And a physician-turned-writer was trying to peddle stories geared to the relatively new genre of detective fiction. As it turned out, one hand fed the other. Sir Arthur Conan Doyle used the budding new field of forensics as a vehicle for his Sherlock Holmes stories, written between 1887 and 1915. The immense appeal of his writing, in turn, fired the imagination not only of the public but also of law enforcement professionals who increasingly turned their attention to newly developing forensic techniques including, but by no means only, fingerprinting.

THE WRITE STUFF
Conan Doyle first portrayed Holmes using science to track down criminals at the same time as Galton was writing his comprehensive works on classification of fingerprints and years before Scotland Yard adopted fingerprint identification in 1900. Knowledge of biology, photography, blood chemistry, botany, and geology were among the scientific tools that Holmes relied on to unravel the mysteries that his creator placed

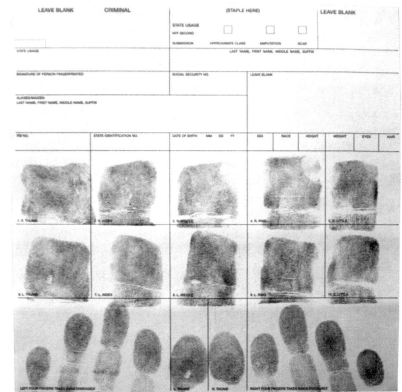

Left: Sherlock Holmes caught not only the imagination of the public but of law enforcement's as well. Right: Authorities use the ten-print card to record a suspect's full set of fingerprints during the booking process; these prints are called "criminal fingerprints" because they are taken as a record of a criminal arrest. Top left: Geology was one of the areas in which Holmes excelled.

before him. Forensics experts and historians today credit the popularity of the Sherlock Holmes stories for generating much of the initial awareness and acceptance of applied science by law enforcement.

Although Conan Doyle's influence was the most profound, other authors—Edgar Allan Poe among them—also helped popularize scientific detective work. It is not an exaggeration to say that fiction helped turn forensics into fact. Academicians have delved into this aspect in the development of the forensic sciences. Ronald R. Thomas, an English professor at Trinity College in Hartford, Connecticut, recognized its importance in an acclaimed book published by the Cambridge University Press. Setting the tone for his book, which is scholarly in its presentation, Thomas asserts that "the history of detective fiction is deeply implicated in the history of forensic technology."

American Edgar Allan Poe was one of the nineteenth-century writers who launched the field of detective fiction.

MEDIA SATURATION

A feedback loop between the practice of forensics in law enforcement and its popularization by the media continues today, possibly even stronger than ever before. Authors turn out a torrent of forensics-themed books, factual-based ones as well as fictional.

It can seem that few sensational homicides, rapes, or kidnappings are reported by television news without commentary by forensics experts. Programming that features forensic investigation has become a staple for television executives seeking to sweep high audience ratings.

AN "ELEMENTARY" SCIENCE

The popularity of forensics today reflects the reality that science has truly become integral to successful police work and gains in importance as research opens new applications of scientific disciplines to law enforcement. The Sherlock Holmes stories, on the other hand, actually anticipated what lay in store for forensics. French police scientist Edmond Locard, founder of the first crime laboratory, is said to have suggested to his students that they read the Holmes stories for guidance. Sherlock Holmes applied the scientific method, expounded earlier by Sir Isaac Newton, to solve crimes. Holmes observed, formulated a hypothesis, experimented to prove whether it was true or false, and arrived at his conclusion.

Dr. Watson gave Holmes high marks in botany, and particularly of poisonous plants such as belladonna, shown above.

DR. WATSON GRADES SHERLOCK HOLMES

In the short story "A Study in Scarlet," in which Sherlock Holmes first appeared, Dr. John Watson offered an evaluation of Holmes's knowledge of the arts and sciences. Watson compiled a list, Sherlock Holmes—his limits, giving his opinion of where Holmes's understanding and ability lay. He flunked the detective in his knowledge of philosophy and astronomy, considering it "nil." Likewise, his knowledge of literature. His grasp of politics was "feeble." Botany was another story. Holmes was sharp on poisons generally, including belladonna and opium. As for geology, Watson gave Holmes credit for distinguishing between different soils and their origins at a glance. His knowledge of chemistry, Watson noted, was "profound," of anatomy, "accurate, but unsystematic," and, the doctor noted, Holmes "plays the violin well."

Conan Doyle as a Detective

A significant number of scholars who analyze the works of great literary figures are of the opinion that Sir Arthur Conan Doyle based the character of Sherlock Holmes largely on Dr. Joseph Bell, a ground-breaking forensics scientist who lectured at the University of Edinburgh's medical school when the author was a student there. In Holmes, however, there was more than a little of Conan Doyle himself. Both Holmes and Conan Doyle were adept at a science not of laboratories, observation, and experimentation: the "sweet science" of boxing. Holmes also practiced Japanese martial arts long before there was a karate or kung fu studio on virtually every street corner. Conan Doyle, moreover, had a bit of a detective in himself.

WAS HE GUILTY?
Bell had taught Conan Doyle to be a keen observer and to link observations with conclusions. Bell was reportedly so adept at the skill of observation that he supposedly could pick a sailor out of a crowd just by noting the way he walked. Applying the lessons of his mentor to a real-life cause, Conan Doyle used a series of remarkable observations,

Above: There was much of his creator in the Holmes character, including a skill at boxing. Top left: To closely examine evidence, Holmes relied on a magnifying glass, which has become a symbol of the character.

along with his medical training as an ophthalmologist, to conduct an investigatory crusade that righted a sensational miscarriage of justice that had been tainted by more than a touch of racism.

In a style worthy of the detective he created, Conan Doyle approached the case with the fervor of a huntsman, his quarry being the truth. George Edalji, a young lawyer with an Indian father and an English mother, had been convicted of mutilating livestock, leaving them to die, and then writing letters to police threatening to do the same to young girls. Edalji went to jail but not everyone was convinced that he was the culprit. A petition was organized to protest

Sir Arthur Conan Doyle (1859–1930)

his conviction and to press for his release. After serving three years of his seven-year sentence, Edalji was indeed freed, although no reason was given for his release, his name was not cleared, and he remained disbarred from practicing law.

A DOGGED INVESTIGATION

Hearing of the case, Conan Doyle felt compelled to act and instigated his own investigation into the matter in 1906. First, he examined the evidence gathered by police at the lawyer's home. He found that a razor was stained not, as police claimed, with blood but was rusted. Spots on Edalji's clothing turned out to be food stains, not fluids from a horse that had been slashed. Mud on boots was of a different composition than that from the field in which the horse had been found. The killings and letters continued after Edalji was prosecuted. When Conan Doyle met Edalji in person, his suspicions of innocence were confirmed. Edalji was able to see lettering clearly enough to read only when he held a document very close to his eyes. He was myopic, with severely impaired vision that would have made venturing into pastures and fields at night hazardous. When police ignored his findings, Conan Doyle began writing articles on the case, which were published in the *Daily Telegraph*. This time he caught the attention of the British government and because at that time there was no procedure for a retrial, a

Sir Arthur Conan Doyle did his share of real-life detective work when he defended a young lawyer he thought wrongfully convicted of a heinous crime. After closely examining a "bloody" straight blade collected by police to convict George Edalji, Conan Doyle determined that the blade in question was stained with rust.

private committee was formed to consider the matter. In the spring of 1907 the committee ruled Edalji innocent of the mutilations, but still found him guilty of writing the anonymous letters. Edalji was finally vindicated and he returned to his legal practice.

THE FINAL PIECE OF THE PUZZLE

Conan Doyle was not finished, however. Probing further, he examined the threatening letters. A difference in the literacy level between several of them indicated to him that two people had written them. Conan Doyle eventually found a suspect, a troubled boy who had attended the same grammar school as Edalji. The boy had eventually been trained as a butcher and went to work in a

slaughterhouse. Neither Conan Doyle nor the legal authorities, however, ever pursued the lead further. Nevertheless, Conan Doyle's investigation and the forensics involved were worthy of Sherlock Holmes at his best.

Although Edalji's muddy boots were used as evidence in the original trial against him, upon reexamination, Conan Doyle definitively determined that the composition of soil samples from the boots did not match the soil composition at the crime scene.

Forensics Fiction

Like Sir Arthur Conan Doyle, who brought scientific knowledge gained while studying medicine under forensics expert Joseph Bell, several modern-day best-selling fiction crime writers have backgrounds in forensics or related fields. Poison was an instrument of homicide in half of the books written by Agatha Christie, the British-born undisputed grand dame of detective fiction and probably the second woman to ever put her name on a major crime novel. Her reliance on poisons seems to be a case of Christie doing what comes naturally, for she worked for a time as a pharmacist. The creator of fastidious Belgian detective Hercule Poirot and British busybody Miss Marple, who still solve mysteries on television today, Christie carefully studied the medical implications of poisons and even worked in a hospital to further her knowledge of forensics.

A NOVEL CONCEPT

Patricia Cornwell, author of a series of novels including *Postmortem*, *Body of Evidence*, and *All that Remains*, based on the adventures and forensic sleuthing of Virginia medical examiner Kay Scarpetta, was a prize-winning crime reporter for the *Charlotte Observer* in North Carolina, and then worked in the office of the Virginia medical examiner. There, Cornwell served as a technical writer and computer analyst. Given Cornwell's background and the nature of her heroine's job, her novels are laced with forensic science, particularly medically related fields. Forensic deductions play a major role in solving the criminal investigations in which Scarpetta becomes involved. Like many other fictitious forensics scientists, Scarpetta is more likely to confront criminals face to face than real-life medical examiners. In one book, *The Body Farm*, which refers to the actual forensics research facility at the University of Tennessee, Scarpetta dispatches the killer with a pump-action shotgun, an affair described in gory detail.

SKELETON KEYS

Body Farm founder Bill Bass, teaming up with journalist and filmmaker Jon Jefferson, who together wrote the nonfiction *Death's Acre*, produced their first work of forensic fiction in 2006. *Carved in Bone* features forensic expert Dr. Bill Brockton, an obvious fictional version of Bass. In the book, Brockton solves the years-old murder of a woman whose body is found in an

Left: Renowned mystery writer Agatha Christie had a science background, as do many writers of the genre. Right: Bill Bass and his writing partner, Jon Jefferson, made the setting of their latest forensics fiction his home turf of Tennessee. Top left: Former pharmacist Christie used poison as a killing device in half of her mysteries.

Appalachian cave, where natural chemical processes have turned it into a mummy. *Carved in Bone* provides accounts of forensic anthropology and other sciences at work with the accuracy of one who actually practices them.

NATURAL PROGRESSION

When it comes to crime fiction and thriller writers with scientific and law enforcement credentials, few writers have the background of Ken Goddard. His books, such as *First Evidence* and *Outer Perimeter,* focus on United States Fish and Wildlife Service law enforcement agents and feature villains ranging from international terrorists to backwoods outlaws. Using the service's law enforcement efforts as a vehicle is natural because Goddard heads the agency's wildlife forensics laboratory, besides being a deputy sheriff and running two crime labs in California. Although Goddard does not use real crimes as models for his stories, he does make use of actual crime scene investigation situations, some from his own experience, to add credibility to his work. Bass and Goddard are among the better-known forensic scientists who have turned their professional experience into source material for novels. Many other forensic scientists have tried their hand at fiction, with varying success, in a market that is driven by a voracious public appetite for stories that bring readers into the realm of forensics investigation.

Best-selling author of The Body Farm, *Patricia Cornwell, shakes hands with Dr. William Bass, founder of the real-life Body Farm.*

Media Superstars

In the slang of television broadcasting, the use of the term "talking head" has been extended from describing a medium camera shot that holds on a person's head and shoulders during an interview to an irreverent reference to pundits and other media personalities who comment on issues in the news. Although they often appear on broadcast networks, talking heads are also a staple of cable news and news commentary, especially on venues offering this sort of programming 24 hours a day. The talking heads offer their insights into politics, entertainment, celebrity scandals, foreign affairs, warfare, and, ever so frequently, the forensics aspects of sensational criminal investigations.

HOUSEHOLD NAMES

There are several forensics pundits who appear so often on television that they have become media superstars. HBO, for example, has aired several specials featuring Dr. Michael M. Baden, a pathologist who was chief medical examiner of New York City and codirects the New York State Police Medicolegal Investigation Unit.

By and large the forensic pundits who appear on television most frequently possess not only expertise in their fields but also considerable presence on camera. Dr. Henry Lee, arguably the dean of them all, is in great demand as a public speaker, not just for his insights into major crimes and the way they are investigated, but also for his witty manner of presentation. Lee has the ability to sweeten accounts most grisly with occasional dashes of black humor.

Above: Dr. Henry Lee talks about a murder suspect's sneakers during a cross examination in court in Durham, North Carolina. Top left: High-profile trials have led law enforcement officials to rethink how they handle evidence recovery.

Lee's personal story itself is fascinating. Born in China, he graduated in 1960 from the Taiwan Central Police College with a degree in police science and rose to the level of captain in the Taiwan Police Department. After immigrating to the United States—arriving there, by his own admission, with little money—he studied forensics and biochemistry, earning a Ph.D. in biochemistry from New York University. Lee went on to establish what became a highly respected forensic science program at Connecticut's University of New Haven, headed the Connecticut State Police crime laboratory, and then became that state's Commissioner of Public Safety.

THE SIMPSON TRIAL

Like Baden, Lee has been the star of his own television show, hosting a series about his work on Court TV. Lee and Baden

have kept the public eye focused on themselves largely by their roles as forensics consultants in high-profile cases. Lee and Baden both received extensive media coverage when they testified as expert witnesses for the defense in the 1995 trial of O. J. Simpson for the murder of his wife, Nicole, and Ron Goldman. Their testimony was part of the defense strategy challenging the integrity of forensics evidence collected at the crime scene. In the eyes of many forensics experts, mishandling of evidence in the Simpson case led law enforcement to rethink the techniques with which evidence is handled and passed through the chain of custody. The Simpson case itself, which held the attention of the world because of the defendant's celebrity status, is one of the factors that kindled public interest in forensics as a whole. It was in the aftermath of

the case that television programs and books about forensics began to flood the market and the appearance of forensics experts to comment on major crimes became a must-have for many television news outlets.

O. J. Simpson won the criminal trial that captured a huge television audience, but lost the less-covered civil case.

Left: Former Los Angeles police detective Mark Fuhrman points to a glove found near the blood-smeared walkway near the scene of Nicole Brown Simpson's murder. Right: Dr. Michael Baden testifies during the trial of former New Jersey Net Jayson Williams.

Forensics on Television

In the history of television, the beginning of the twenty-first century may be remembered as the time when the men and women in lab coats took over crime dramas. One way to say it is that the tough guy was replaced by the smart guys, female gender included. Hard-as-nails Sam Spade and taciturn, poker-faced Jack Webb of *Dragnet* are no longer the faces of criminal investigation. Instead, the main character in a television crime drama is more likely to be a Ph.D. educated at an Ivy League university who is more adept at biochemistry or materials analysis than in running down and slugging it out with a fleeing criminal. Forensics scientists are the new heroes and heroines of television crime shows.

A BREAKTHROUGH SHOW

The drama that was in the first wave of a flood of forensic-based crime dramas, reality-based shows, and documentaries that are offered to television viewers was *CSI: Crime Scene Investigation*, which premiered

Top: Forensics-themed shows dominate TV's prime time. Bottom: An actor portraying a corpse waits during his costar's makeup call on the set of CSI: Miami. Top left: Fictional portrayals of forensic investigations use tools similar to real life, but with higher success rate.

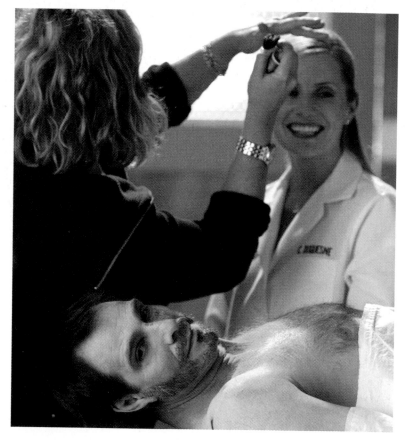

on CBS in the fall of 2000. Set in Las Vegas, the show introduced a team of forensic investigators in a crime scene investigation unit and followed their efforts to use science to solve crimes based upon evidence left at the scene. It quickly spawned spin-offs, based in Miami and New York City and became a ratings phenomenon. A host of other shows keyed to forensics followed on broadcast television and cable. Among them: *Crossing Jordan* on NBC, *Cold Case Files* on A&E, and *Bones* on Fox.

HIGH PROFILE CASES

Many television executives credit the endless coverage of the O. J. Simpson trial for the explosive increase in the popularity of forensic science. The focus of the trial on evidence recovery—on footprints, gloves, and expensive shoes—made crime scene investigation and attendant forensics fields a "glamour" business. The O. J. trial also made news producers more attentive to covering forensics aspects of other major crimes, such as the JonBenet Ramsey murder and the BTK serial killer case. Increased news coverage in turn inspired more new forensic shows. Although the mass popularity of forensics-based programming is phenomenal, the programming itself is not. During the 1970s, *Quincy M.E.* had a long run as a series about a crime-fighting medical examiner, but it was tame compared to today's shows in terms of details, both about science and about gore.

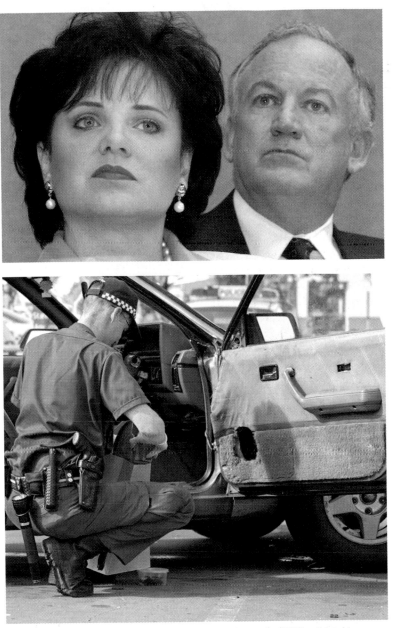

Top: Patricia and John Ramsey during a press conference in 2000, after taking lie-detector tests concerning the death of their daughter JonBenet. Bottom: Although the public might be better educated about evidence recovery than they were a decade earlier, the task of evidence recovery is best left to an experienced technician.

Cracker, produced in the United Kingdom and widely aired around the world, featured a criminal psychologist, Eddie Fitzgerald, who solved crimes despite his womanizing, heavy drinking, addictive gambling, and jaded view of life.

The CSI Effect

Above: CSI: Crime Scene Investigation *cast member Gary Dourdan stands in the spotlight at a Las Vegas, Nevada, media affair. Top left: You might need your cable guide to tell you if what you are watching is real or just a fictional depiction.*

It was not the sort of topic generally discussed at symposia held during the 2005 annual meeting of the American Association for the Advancement of Science. The subject was a television crime show. More specifically, the impact of the *CSI* shows broadcast by CBS on the public perception of forensics and how it affects the attitudes of juries towards evidence offered in court. The title of the symposium was "The CSI Effect: Forensic Science in the Public Imagination." During the symposium, real-life forensics experts opined that their fictional counterparts in television dramas contribute to unrealistic expectations about the accuracy and inclusiveness of forensics that has an influence all the way into the jury room.

FICTION VERSUS REALITY

The term "CSI Effect" has become one that has been regularly used in the scientific, legal, and law enforcement communities to describe how television has shaped popular attitudes on the role and importance of forensics. Concern about the CSI Effect has been expressed by forensics experts from academicians at institutions such as the Forensic Pathology Unit at Leicester University and the

Centre for Forensic Science at Strathclyde University in the United Kingdom, and National Clearinghouse for Science, Technology and the Law at the Stetson University College of Law in the United States. Both prosecutors and defense attorneys worry that a public image of forensics evidence as infallible can skew the decisions of juries. Victims of crimes and their families now demand that police provide forensic evidence even if, as often can be the case, it is not available. Some critics of forensic television dramas say that these shows never present

Crime scene investigator demonstrates his process for lifting fingerprints.

A jury acquitted a man of drug possession because police had not finger-printed the package of crack cocaine.

cases in which forensic evidence has little or no impact or is faulty. Another criticism is that some of the technologies used by television science sleuths is still highly experimental and unavailable in real-life situations. "I wish we had what they have," remarks one veteran law enforcement officer. "We'd solve a lot more crimes."

AMATEUR SLEUTHS

There is substantial evidence that certain juries do seem to have been influenced by what they have seen on television. In Arizona, jurors asked for a DNA test of a bloody coat offered as evidence even though the defendant admitted to being at the crime scene. A Virginia jury acquitted a man charged with drug possession after tossing a large amount of crack cocaine out of his car when stopped by police because the authorities had never checked the drugs for fingerprints. In May 2005, the *National Law Journal* reported that some attorneys believe that a diet of television crime shows

There are some attorneys and others in the judicial system who have become wary of jurors who spend much of their time watching fictional forensics-themed television shows and therefore expect forensic evidence to be relevant in all criminal cases.

has caused some misguided jurors to try to investigate and solve cases on their own, in defiance of orders from the bench.

UNREALISTIC EXPECTATIONS?

Some law enforcement officials complain that many jurors now expect virtually every item collected at a crime scene to be forensically tested, even if testing is unnecessary. Some prosecutors even interview prospective jurors on their television viewing habits and try to assess if they are influenced by forensics shows. Prosecutors have also spent time explaining to juries

that not everything they see on forensics-themed television programs can be applied to real-life criminal investigations.

Much of the evidence linking television programming to juries gone astray is anecdotal. One study by Kimberlianne Podlas, an attorney who teaches media studies at the University of North Carolina, argues to the contrary. Podlas presented a hypothetical criminal case to a group of subjects who had been divided by those who watched the *CSI* shows and those who did not. The verdict: Television had no impact on whether they decided for guilt or innocence.

Pop Science, Real Science

In 1999, when Max Houck at West Virginia University initiated the graduate-level Forensic and Investigative Sciences Program, a mere nine students were graduated from it. Just six years later, the number of students enrolled in the program totaled 400. Television shows that feature forensic investigation may be regarded in some quarters as pop science but they have fostered a remarkable growth in the number of students interested in studying hard sciences, in this case, science applied to forensics.

FORENSICS BOOM

There are more than 100 colleges and universities in the United States—and hundreds more elsewhere in the world—that offer programs in the forensic sciences. These include schools ranging from those that offer two-year associate's degrees to Ph.D. programs. Some students major in forensics while others opt for majors in sciences such as biochemistry and chemistry with minors in forensics, or pursue forensics, legal, or law enforcement studies after attaining a bachelor's degree in science. Many schools, such as the University of West Virginia, have programs that are relatively new. Others, notably the University of

Top: Investigators at a mock grave site learn new skills at a seminar offered at the International Association for Identification. Bottom: A girl examines hair strands from both humans and animals during a forensics program designed for children. Top left: Forensics's popularity has led to a rise in students of hard science.

New Haven in Connecticut, have offered programs in forensics for decades, in the case of New Haven, for more than 30 years.

There are myriad examples. The University of Rhode Island, for example, offers a minor in forensic science. Students study fields such as criminalistics while majoring in other fields, mostly the sciences or engineering, but in other areas of concentration as well. West Virginia University has a forensic science and investigative science major that provides a background in physical and biological sciences with courses such as fingerprinting, evidence recovery, and other pure forensics subjects.

MASS APPEAL

Some colleges with education majors are offering courses that prepare students for teaching forensics at secondary and elementary school levels. It is a reflection of the fact that many high schools, junior high schools, and even elementary schools are using forensics classes to interest students in science. As the *Wall Street Journal* noted in a February 2002 article "Gore Curriculum," the courses are designed to compete with fast-paced, reality-based entertainment for the attention of youngsters weaned on such fare. The National Science Teachers has noted that forensics courses can be vehicles for motivating student interest in science.

TOO MUCH, TOO SOON?

High school students have gone on trips to mock crime scenes,

Top: Students in forensics programs see sights such as maggots feeding on human flesh. Bottom: Some critics argue that school mock crime scenes are too realistic, and that children and teenagers should not be exposed to real corpses and the like.

complete with mannequins representing dead bodies, replicas of severed limbs, and weapons coated with artificial blood. They have visited morgues, viewed real corpses, and observed maggots—those important indicators of the time of death—feeding on human remains. Students also have reenacted crimes. Teachers have even acted the role of victims.

Some educators say the attempt to simulate realism may be excessive and even harmful.

Young people, they insist, are already exposed to far too much violence in so many other venues. They question whether students should become too involved with some of the morbid aspects of forensics by classes that take realism a step too far. Indeed, some classes can be very realistic. One teacher, reported the *Wall Street Journal*, corresponded with imprisoned cult murderer Charles Manson and had students read their letters as part of class.

GLOSSARY

A

ARCH A type of characteristic ridge pattern in a fingerprint.

AUTOPSY Complete examination of a body by a physician to determine cause of death.

B

BALLISTICS The study of firearms and ammunition characteristics.

BLOOD SPATTER The shape and pattern of blood that has emanated from a wound and is deposited on surfaces.

BLOOD SPATTER ANALYSIS Examining blood spatters to determine circumstances of how a wound was created.

C

CALIBER Internal diameter of a gun barrel. Applied to bullets that fit it.

CHROMOSOMES Threadlike bodies in the nucleus of a cell that carries genes. Humans usually have 23 pairs each.

COMPARISON MICROSCOPE Two microscopes linked by an optical bridge with a split screen allowing both fields of view to be seen simultaneously.

CSI EFFECT The phenomenon in which the popularity of forensics-themed television shows have raised crime victims' and jury members' expectations of forensic evidence, DNA testing, and crime scene investigation to the level depicted in these shows.

D

DACTYLOSCOPY The technique of developing and identifying fingerprints.

DNA Deoxyribonucleic acid. The basic genetic material in cells constructed of a double helix and found in chromosomes.

DNA PROFILING Testing to identify specific DNA patterns or types.

DOUBLE ACTION Gun action in which a trigger pull cocks a gun for firing a next round.

DOUBLE HELIX Molecular structure of DNA discovered in 1953.

E

ELECTRON MICROSCOPE A microscope that creates an image from electrons emitted by the subject under view.

EVIDENCE RECOVERY Detecting and removing evidence at and from a crime scene.

F

FINGERPRINTS Skin surface pattern of ridges on palm side of fingers believed unique to each person.

FORENSIC SCIENCE Science applied to the legal system.

I

INFRARED Electromagnetic radiation, sensed as heat and invisible to the eye with a wavelength slightly longer than visible light.

G

GAS CHROMATOGRAPHY Analyzing a material by breaking it down into its components through heating it into a gas and tracking the migration routes of each substance it contains.

GUNSHOT RESIDUE Unburned powder left on a target and surrounding area after a bullet is fired.

I

IMPRESSION EVIDENCE Footprints, tire marks, and anything else that creates an impression that can be used as evidence of a crime.

J

JURISDICTION Authority of a governmental entity over individuals or legal matters within a defined geographical area.

L

LATENT Not visible to the eye alone.

LIVOR MORTIS Discoloration of body after death caused by settling of red blood cells by gravity.

M

MASS SPECTROMETRY Identifying elements in a compound by vaporizing it with an electric discharge and subjecting it to electron beam.

MITROCHONDRIAL DNA A type of DNA found outside the cell's nucleus in units called "mitochondria" and inherited along the female line only.

O

ODONTOLOGY The study of teeth.

P

PALYNOLOGY Study of pollens, living and fossil.

PHYSICAL EVIDENCE An object that can establish commission of a crime or link those involved.

POLYMER Long-chained, complex molecule.

R

RIDGE Raised lines in fingerprints.

RIGOR MORTIS Stiffening of body after death.

S

SCANNING ELECTRON MICROSCOPE An electron microscope of great power that can magnify to a factor of 100,000.

SINGLE ACTION Gun action that must be manually cocked for firing.

STRIATIONS Lines on bullet left by rifling of a firearm.

T

TRACE EVIDENCE Small physical evidence that can be collected from a crime scene, such as hairs, fibers, and dust.

U

ULTRAVIOLET Radiation invisible to the eye, having a wavelength shorter than wavelengths of visible light and longer than those of X-rays.

W

WHORLS Fingerprint patterns of ridges completing a full circuit.

FURTHER READING

BOOKS

Baden, Michael M. *Unnatural Death: Confessions of a Medical Examiner.* New York: Ballantine, 1990.

Bass, Dr. Bill, and Jon Jefferson. *Carved in Bone: A Body Farm Novel.* New York: William Morrow, 2006.

———. Death's Acre: *Inside the Legendary Forensic Lab the Body Farm Where the Dead Do Tell Tales.* New York: Berkley Books, 2003.

Cornwell, Patricia. *The Body Farm.* New York: Berkley Books, 1995.

Camenson, Blythe. *Opportunities in Forensic Science Careers.* New York: McGraw-Hill, 2001.

Genge, Ngaire E. *The Forensic Casebook: The Science of Crime Scene Investigation.* New York, Ballantine, 2002.

Kirwin, Barbara. *The Mad, the Bad, and the Innocent: The Criminal Mind on Trial—Tales of a Forensic Psychologist.* New York: HarperTorch, 1998.

Lee, Henry C., and Thomas W. O'Neil. *Cracking More Cases: The Forensic Science of Solving Crimes: The Michael Skakel–Martha Moxley Case, the JonBenet Ramsey Case and Many More.* Amherst, NY: Prometheus Books, 2004.

Lyle, Douglas P. *Forensics for Dummies.* 2004.

Manheim, Mary H. *Bone Lady: Life as a Forensic Anthropologist.* New York: Penguin, 2000.

Mann, Robert, and Miryam Ehrlich Williamson. *Forensic Detective: How I Cracked the World's Toughest Cases.* New York: Ballantine Books, 2006.

Maples, William R., and Michael Browning. *Dead Men Do Tell Tales: The Strange and Fascinating Cases of a Forensic Anthropologist.* New York: Broadway Books, 2001.

Rainis, Kenneth G. *Crime-Solving Science Projects: Forensic Science Experiments.* Berkeley Heights, NJ: Enslow Publishers, 2000.

Ramsland, Katherine M. *Forensic Science of CSI.* New York: Berkley Books, 2001.

———. *The Human Predator: A Historical Chronicle of Serial Murder and Forensic Investigation.* New York: Berkley Books, 2005.

Roach, Mary. *Stiff: The Curious Lives of Human Cadavers.* New York: W. W. Norton & Company, 2004.

Stevens, Serita. *Forensic Nurse: The New Role of the Nurse in Law Enforcement.* New York: Thomas Dunne Books, 2004

Thomas, Ronald R. *Detective Fiction and the Rise of Forensic Science.* New York: Cambridge University Press, 2004.

Ubelaker, Douglas H., and Henry Scammell. *Bones.* New York: HarperCollins, 2000.

Warnock, Bill. *The Dead of Winter: How Battlefield Investigators, WWII Veterans, and Forensic Scientists Solved the Mystery of the Bulge's Lost Soldiers.* New York: Chamberlain Bros., 2005.

WEB SITES

American Academy of Forensics Sciences
www.aafs.org
Discipline and career information.

Court TV Crime Library
www.crimelibrary.com
Nonfiction feature stories on major crimes, criminals,
trials, forensics, and criminal profiling.

Federal Bureau of Investigation
www.fbi.gov.
United States government agency.

Georgia Bureau of Investigation
www.state.ga.us/gbi
Investigative unit of the State of Georgia.

Los Angeles County Department of Coroner
coroner.co.la.ca.us
Information on coroner operations.

Scientific Testimony: An Online Journal
www.scientific.org
Forensic profiling information.

United States Fish and Wildlife Service
Forensics Laboratory
www.lab.fws.gov
The only crime lab devoted entirely to wildlife.

AT THE SMITHSONIAN

Scholars of forensics and research scientists from around the world regularly visit the valuable collections and resources relating to physical anthropology at the Smithsonian Institution's National Museum of Natural History.

The roots of Smithsonian Institution contributions to forensic science date back to consultations with Aleš Hrdlička, a pioneer in American physical anthropology. Hrdlička became the first curator of physical anthropology at the National Museum of Natural History in 1903. As early as 1910 Hrdlička reported on a case in Argentina. By the 1930s he was discovered by the FBI in Washington, D.C., and began regular consultation on skeletal cases and other forensic issues until his death in 1943. The vast collections of human remains from around the world, many assembled by Hrdlička, represent valuable resources that contribute to the applied comparative science that is forensic anthropology. Although these collections are not accessible to the general public, research scientists from around the world routinely visit them.

The Federal Bureau of Investigation in Washington, D.C. Since the inception of its crime laboratory, the FBI and the Smithsonian have shared a strong working relationship.

Hrdlička's successor as curator of physical anthropology, T. Dale Stewart became a routine consultant in forensic anthropology for the FBI, reporting on approximately 250 cases for the FBI and others, mostly between 1942, when he took over from Hrdlička as curator, and 1962, when he became the museum's director. Stewart also launched major research projects on forensic issues and published a great deal on forensic topics, including *Essentials of Forensic Anthropology*. When Stewart became the museum director, J. Lawrence Angel was hired as curator of physical anthropology. Although he had little previous experience with forensic applications, Angel immediately assumed responsibility for the FBI consultation and reported on about 565 cases between 1962 and 1986.

In 1977, Angel took a much-needed sabbatical year off to write, and curator Douglas H. Ubelaker assumed the role of consultant with the FBI, a role

Since 1977, National Museum of Natural History curator Douglas H. Ubelaker has consulted on nearly 800 FBI cases.

he continues in today, nearly 30 years later. Over the years, Ubelaker has reported on 767 cases and has conducted extensive research aimed at improving methodology in forensic anthropology. In 2006, Ubelaker continues consultation with FBI headquarters and represents their primary consultant in forensic anthropology. In addition, Douglas Owsley consults with West Virginia and others in forensic anthropology and David Hunt reports on cases for the District of Columbia.

CONTRIBUTIONS TO FORENSICS FROM OTHER DIVISIONS

Although traditionally, most of the input into forensic science at the Smithsonian Institution has centered on physical anthropology within the Department of Anthropology of the Smithsonian Institution's National Museum of Natural History, specialists in other departments are called upon as needed. In particular, the Division of Birds has proven to be of great assistance in the identification of birds involved in aviation accidents. Their vast collections of birds and feathers have enabled identification and have been used in training.

The National Museum of Natural History maintains vast and world-class collections in all of the departments, including such key areas as entomology, botany, mineral sciences, paleobiology, vertebrate zoology, and invertebrate zoology. These collections and the scientists who curate them frequently are called up for the special problems of identification presented by individual forensic cases.

Outside of the Department of Physical Anthropology, the vast collections of birds and feathers held by the Division of Birds is an invaluable resource for forensics investigators looking into aviation accidents involving birds.

INDEX

TWA Flight 800, *22*
typewriter, *99*

U

U.S. Bureau of Customs and
  Border Patrol (BCP), 46
U.S. Centers for Disease
  Control, 122
U.S. Fire Administration, 138
U.S. Fish and Wildlife Services,
  42–43
U.S. Food and Drug
  Administration, *130*
U.S. Office of Strategic
  Services, 145
U.S. Postal Inspection Service,
  47
U.S. Secret Service, 47
United Nations, 82, 183
University of Edinburgh, 9, 76
University of Glasgow, 76
University of New Haven,
  202–203
University of Rhode Island, 203
University of Tennessee
  Forensic Anthropology
  Facility, 78

V

Valla, Lorenzo, 13
Vanzetti, Bartolomeo, 13
venom, 131
veterinary forensic pathology,
  57
victimology, 146

Vollmer, August, 16, 101
Vucetich, Juan, 12

W

Wall Street Journal, 203
war crimes, 82–83
Warner, Mark R., 118
Washington Post, 90
Watson, James, 111
Watson, John, 191
weather, 166–167
Webb, Jack, 198
West Virginia University,
  202–203
Westveer, Arthur E., 131
White, Tim, 179
white-collar crime, 104
Widmark, Erik M. P., 11
Wilhelm, Karl, 100
Williamson, Donna, 117
World Trade Center, *19*,
  184–185
Wu Pu, 6

X

Xi Yuan Ji Lu, 6, 100
X-ray, *72*

Z

Zacchia, Paolo, 8–9, 100

ACKNOWLEDGMENTS & PICTURE CREDITS

ACKNOWLEDGMENTS

The author and publisher gratefully acknowledge Douglas H. Ubelaker of the National Museum of Natural History, Smithsonian Institution; Ellen Nanney, Senior Brand Manager, Smithsonian Business Ventures, and Katie Mann, Smithsonian Business Ventures; Collins Reference executive editor Donna Sanzone, editor Lisa Hacken, and editorial assistant Stephanie Meyers; Hydra Publishing president Sean Moore, publishing director Karen Prince, project editor Lisa Purcell, art director Edwin Kuo, designers Rachel Maloney, Mariel Morris, Gus Yoo, Greg Lum, La Tricia Watford, Erika Lubowicki, editorial director Aaron Murray, editors Marisa Iallonardo, Marcel Brousseau, Molly Morrison, Suzanne Lander, Gail Greiner, Ward Calhoun, and Emily Beekman, copy editors Glenn Novak and Eileen Chetti, picture researcher Ben DeWalt, production manager Sarah Reilly, production director Wayne Ellis, and indexer Jessie Shiers; Chrissy McIntyre of Chrissy McIntyre Research, LLC; Betsy Glick of the Federal Bureau of Investigation; Wendy Glassmire of the National Geographic Society; Harriet Mendlowitz of Photo Researchers, Inc.; Crystal Smith and Beth Mullen of the National Library of Medicine; and Iris Stevens of the Westchester County Film Office.

Ready Reference
100cl SIL 100tr SIL 100br SIL 101bl
JI 101c SS/Rachel B Kretch 101cr
LoC 101br iSP/Creative Studios 102tl
iSP/Christine Balderas 102br SS/Emin
Kuliyev 103tr IO/FogStock, LLC 103br
SS/Yegor A. Mandra 104tr JI 104br
SPL/Mauro Fermariello 105br JI 106tl
SS/Emily2k 106tc SS/photobar 106tr IO/
FogStock, LLC 106cl SS/Jessica Bethke
106c iSP/Jose Gil 106cr SS/Lancelot et
Naelle 106bl Tolstoy/photobank.keiv.
ua 106bc iSP/Aleksander Bolbot 106br
SS/Andrei Orlov 107tl SS/Jamie Wiilson
107tcl JI 107tcr iSP/Ron Hohenhaus 107tr
SS/Rohit Seth 107cl SS/Glen Jenkinson
107ccl JI 107ccr SS/Falk Kienas 107cr
BS 107bl iSP/Yin Chern Ng 107bcl
SS/Alexandru Cristian 107bcr SS/Damian
Herde 107br iSP/Andrezej Tokarski

**Chapter 7: DNA and Microbial
Forensics**
108 iSP/Spectral Design 109t iSP/Rob
Fox 109b JI 110tl iSP/Mark Evans 110cr
iSP/Jim DeLillo 111bl IO/photos.com
Select 111tr PR/SPL 112tl iSP/SX70
112bl IO/photolibrary.com pty. Ltd. 113tc
iSP/Katherine Garrenson 113cr iSP/Peter
Chen 114tl iSP/Katherine Garrenson
114br iSP/Andrei Tchernov 115cr
iSP/Leonardo Fabbri 115bc iSP/Andrei
Tchernov 116tl iSP/Andrei Tchernov
116br iSP/Stefan Klein 117tr iSP/Floyd
Willis 117bc iSP/Andrei Tchernov 118tl
iSP/Andrei Tchernov 118cr AP/Steve
Helber 119bl iSP/Adam Tinney 119br
iSP/Andrei Tchernov 120tl IO/photos.
com Select 120bl IO/LLC, FogStock
120br IO/LLC, FogStock 121bl iSP/
Mikhail Kondrashov 122tl iSP/Monika
Wisniewska 122c iSP/Bertrand Collet
123t CDC123b CDC

Chapter 8: The Chemistry of Crime
124 JI 125t JI 125b SS/James Davidson
126tl JI 126cr JI 126br JI 127cr SS/Robert
Kyllo 127br PR/Science Source 128tl JI
128cr JI 128br iSP/Carsten Madsen 129tl
SS/Minkia Adamczyk 129br SS/Michael
McCloskey 130tl JI 130bl SIL 130br
IO/FogStock LLC 131tr CDC 132tl
SS/Lyle E. Doberstein 132cr IO/FogStock
LLC 132br SS/Foyik Yevgen 133tl
IO/AbleStock 133bl JI 134tl SS/Adrian
Hughes 134bl SS/Ljupco Smokovski
134br SS/Lincoln Rogers 135tr SS/Girish
Menon 135br JI 136tl JI 136cr iSP/Eliza
Snow 136br SS/Charles Siilvey 137tl JI
137tr SS/Robert Manley 138tl SS/Dale
A Stork 138cr SS/Dale A Stork 138br

JI 139tl JI 139br SS/Jack Dagley
Photography

Chapter 9: Profiling Criminals
140 SS/Vlad Mereuta 141t SS/Tiburon
Studios 141b SS/Oleg Kozlov, Sophy
Kozlova 142tl SS/Robert Mizerek 142br
SS/Jack Dagley Photography 143tr
SS/Jack Dagley Photography 143br
SS/Carolina K Smith 144tl SS/Stephanie
Tougard 144c SIL 144cf SS/Emma
Holmwood 145tl SS/Bill Kennedy 146tl
SS/Cesair 146bl JI 146br SS/Ann Marie
Hughes 147cr SS/Fiorentini Massimo
147br SS/Dale A Stork 148tl iSP/Rasmus
Rasmussen 148cr iSP/Dan Brandenburg
148br IO/DesignPics, Inc. 149tc SS/Andre
Klaassen 150tl iSP/Constantine the
Second 150bl AP/Travis Heying, Pool
150br AP/Charles Krupa 151tr
SS/Grant Blakeman 152tl JI 152bl
SS/Pichugin Dmitry 152br iSP/Matt
Knannlein 153tl JI 153tr iSP/Stephen
Mulcahey 153br SS/Varina Hinkle 154tl
SS/Eon Alers 154bl SS/Aleksandar
Bracinac 154cf SS/Harald Hoiland
Tjostheim 155tl SS/Bob Denelzen 155tr
SS/Joseph 155br SS/Tina Rencelj

Chapter 10: Clues from Nature
156 SS/Riccardo Bastianello 157t JI
157b JI 158tl SS/Vera Bogaerts 158bl
IO/Vstock, LLC 158br JI 159t JI 159b
iSP/Alessandro Terni 160tl JI 160tr JI
160br JI 161tl SS/Denise Sirois 161tr
SS/Wolfgang Scheitinger 162tl JI 162bl
iSP/Christian Anthony 162br JI 163t
SS/Gary Fowler 163b SS/Svetlana
Tikhonova 164tl iSP/Gerry Okimi 164tr
BS/Romko 164br iSP/Clayton Hansen
165t JI 165c iSP/Jostein Hauge 165b
SS/Jim Orr 166tl iSP/Oleg Prikhodko
166tr iSP/Ahmed Hussam 166br JI 167tl
iSP/Paulus Rusyanto 167tr iSP/Cristian
Matei 167br JI 168tl JI 168tc iSP/royrak
168tr SS/Gary Unwin 168b PR/ Claude
Nuridsany and Marie Perennou 169 SPL/
Claude Nuridsany and Marie Perennou
170tl BS 170bl PR/John Mitchell 170br
iSP/Rebecca Ellis 171t iSP/Josef Kubicek
171b SPL/Gary Meszaros

Chapter 11: Digging into Mysteries
172 AP/Amel Emric, Stringer 173t SS/Alex
Balako 173b SS/Alex Balako 174tl SS/Pepe
Ramirez 174bl SS/Vladimir Pomortzeff
174br BS/WizData 175tl iSP/BMPix
175tr LoC 175br AP/Stringer Ho 176tl
SS/Anton Poluektov 176bl AP/South
Tyrol Museum of Archaeology/Augustin
Ochsenreiter 176br CDC 177tr National

Geographic Image Collection/Stephen
Alvarez 178tl iSP/Knud Nielsen 178br
PR/Tom McHugh 179tl iSP/Fanelie
Rosler 179br AP/ Frank Franklin II 180tl
iSP/Martina Berg 180bl SS/Thierry
Maffeis 181br AP/Gene J Puskar 181tr
iSP/William Fawcett 182tl Public
Domain 182bl AP/Silvia Izquierdo 182br
SS/Carsten Erler 183tl SS/Bartlomiej K.
Kwieciszewski 183 iSP/Andrei Tchernov
184tl SS/Carlos Sanchez Pereyra 184bl
AP/Suzanne Plunket 185tr iSP/Amanda
Potter 185br iSP/Andrei Tchernov 186tl
SS/Pieter Janssen 186tr SS/Paul Prescott
186br iSP/Grace Tan 187tr AP/David
Longstreath 187cr SS/Steven Collins

Chapter 12: Forensics, Fact and Fiction
188fp LoC 189t SS/VM 189b IO/
photolibrary.com pty. Ltd. 190tl SS/Cindy
Hughes 190bl JI 190br SS/Kevin L
Chesson 191bl JI 191tr SS/Michael
Ledray 192tl Wikimedia 192br LoC 192bc
LoC 193tr SS/Justine Olson 193bl SS/
Marilyn Barbone 194tl SS/Alan Egginton
194bl AP 194bc AP/Sayyid Azim 195bc
AP/Wade Payne 196tl SS/John Weise
196br AP/Bill Wilcox 197bl AP/LAPD
197cr AP/Richard Drew 197br AP/Ed
Pagliarini 198tl JI 198cr SS/Steven Pepple
198br AP/Mark J. Terrill 199tr AP/Ric
Feld 199cr SS/Stuart Elflett 200tl SS/Isaiah
Shook 200cr AP/John Locher 200br
AP/Derik Holtmann 201tl Wikimedia
201tr PR/David R Frazier 202tl SS/Emin
Kuliyev 202cr AP/Joe Cavaretta 202br
AP/Wade Payne 203tr iSP/Perry Wang
203cr iSP/Dan Brandenburg

At the Smithsonian
208t Smithsonian Photographic Services/
Dane A. Penland 208b BS/Chris DoDutch
209t Photograph by James Di Loreto,
Smithsonian Institution 209b SS/iwka